# UNDERSTANDING MASS SPECTRA— A BASIC APPROACH

# UNDERSTANDING MASS SPECTRA— A BASIC APPROACH

**R. MARTIN SMITH**
Wisconsin Department of Justice
Crime Laboratories

with

**KENNETH L. BUSCH,** Technical Editor
School of Chemistry and Biochemistry
Georgia Institute of Technology

A Wiley-Interscience Publication

**JOHN WILEY & SONS, INC.**

New York • Chichester • Weinheim • Brisbane • Singapore • Toronto

*Library of Congress Cataloging-in-Publication Data:*
Smith, R. Martin.
    Understanding mass spectra : a basic approach / by R. Martin Smith
  : with Kenneth L. Busch.
      p.  cm.
    Includes index.
    ISBN 0-471-29704-6 (alk. paper)
    1. Mass spectrometry.  I. Busch, Kenneth L.  II. Title.
QD96.M3S65  1999
543′.0873—dc21                        98-18136
                                                CIP

Printed in the United States of America.

10  9  8  7  6  5  4  3  2  1

To
All of My Students

# CONTENTS

# FOREWORD

During a short course on mass spectrometry, a discussion of a research project, or a spirited discussion on the intricacies of spectral interpretation, the undersigned is likely to utter the phrase "The ions know what they're doing." This is an acknowledgment that the tools of mass spectrometry, as complex and sophisticated as they are, can sometimes seem like sledgehammers in tapping out detailed outlines of ionic behavior.

On the surface, the situation would seem ideal. We make molecular ions in a picosecond from the gas-phase neutral molecule, and then, in the complete absence of solvent and other mediating matrix effects, we examine the dissociations of those ions. What could be more simple? To interpret electron ionization mass spectra, we need only reassemble the parts and pieces of the ionic puzzle to deduce the structure of the original neutral molecule. As with a jigsaw puzzle, success should be guaranteed with time and perseverance.

Instead, we give up too easily. We leave ions of unknown origin unlabeled in our mass spectra, preferring instead to label only the ions that can be easily assigned. Maybe we are unsure of the purity of our sample or of our own skills in preventing the introduction of contaminants into the system. Maybe we have run out of time or interest. Perhaps we have simply run out of understanding.

One could not wish for a greater challenge. Rearrangements in molecular ions lead to subtle structural information and differentiations that otherwise could not be obtained. Measuring ionic dissociations as a function of ion internal energy provides an understanding of the reaction surface. Isotopic labeling experiments are clues to deducing dissociation mechanisms, but also are central to measurements of kinetic isotope effects. Key to interpretive acumen is the motivation to rise to these challenges. But we have to start somewhere, and we should start with straightforward and well-explained examples of interpretive strategy.

Such is the value of the present text. Most current monographs in mass spectrometry focus on ionization methods, new experiments, or novel instrumentation in the field. But electron ionization is still the most widely used ionization source, and interpretation of electron ionization mass spectra remains central to the deduction of molecular structure, and underlies the more advanced areas of ion energetics. This book is the first new text in many years to provide support for the development of interpretive skills and reflects R.M. Smith's practical real-world experience in the interpretation of mass spectral data, often from very complex sample mixtures.

Hallmarks of the current text include its comprehensive discussion of isotopic abundances and how they are used with discernment in the interpretation of electron ionization mass spectra; a coherent and usable approach to electron accounting and energy levels in molecular orbitals; and multiple examples based on progressively higher mass examples with more complex structures. By using mass spectra from examples that are structurally related, both the commonalities and the differences are easily seen by the reader. The presumption here is that each of the ions in a mass spectrum is worthy of attention. Since the ions know what they are doing, so will we. Following through the logical discussion that accompanies each example, the reader is brought through the strategies of spectral interpretation that are useful in a functioning analytical laboratory.

This work presents material at a level appropriate for readers who encounter it for the first time, with descriptions that are clear and complete. The text includes material so that the reader understands basic instrumentation used for the acquisition of mass spectra. Isotopes and their impact on the appearance of the mass spectrum and its interpretation are discussed at length, a simple, central concept so often misunderstood by those outside the field. The impact of the computer data system and modern tools such as library searches are discussed with a practical point of view, and always with the approach that the final responsibility for correct identification and accurate quantitation rests with the analyst and the instrument operator.

KENNETH L. BUSCH

*School of Chemistry and Biochemistry*
*Georgia Institute of Technology*

# PREFACE

Gas chromatography/mass spectrometry (GC/MS) has become an indispensable tool in analytical organic laboratories over the past two decades. Businesses and government facilities specializing in the analysis of pesticide residues, environmental pollutants, flavors and fragrances, therapeutic drugs and drugs of abuse, forensic arson residues, and scores of other sample types, all use GC/MS routinely to screen, quantitate, and unambiguously identify organic compounds of interest. Research laboratories depend on it to separate complex reaction and natural product mixtures and to provide structural information about newly isolated or synthesized compounds.

My career in forensic chemistry began with temperamental GC/MS units that filled nearly half a room and demanded user input at virtually every step of the analysis. Since that time hardware has been streamlined so that maintenance and downtime is nearly nonexistent, and data manipulation is so sophisticated that in-depth knowledge about mass spectrometry by the instrument user seems to have been rendered almost superfluous. Indeed, instrument manufacturers tout the fact that essentially anyone can use their equipment with little or no prior training. In our laboratory, each GC/MS unit is preprogrammed so that the user need only pick from a menu on the computer monitor a method to suit his or her sample type, inject an appropriate amount of sample, and press START. A computer controls the GC temperature program, collects mass spectral data over all or part of the chromatographic run, selects chromatographic peaks for spectral presentation, performs library searches on the selected spectra, identifies individual components, and prints out all of this information in a form suitable to our laboratory's needs. The analyst literally needs only return several minutes later and pick up the pages with the "answer" printed on them. With automatic samplers, large numbers of samples are run sequentially with very little operator input.

This procedure works well, but only if certain assumptions hold. Primary among these are that the number of known compounds likely to be encountered during analysis is relatively limited and that the user is not particularly concerned about unknowns that fall outside this group. Underlying this further is the assumption that little, if any, of the data generated by the user's instrument or of the spectral libraries or pertinent literature spectra used as references will need critical evaluation. In my opinion at least one person on each staff (preferably as many as possible) should be proficient enough in mass spectral interpretation to perform this critical evaluation. Without it, the probability of false-positive identifications, overlooked unknowns, and missed research opportunities is high.

Interpretation of mass spectra has traditionally not been an easy subject to approach, and the field continues to become more and more complex. During the 20 years I have taught mass spectral interpretation to forensic chemists and chemists in the health and environmental fields (most of whom have bachelor's degrees in chemistry or a related area), I have found that a large barrier to understanding mass spectral fragmentations lies in approaches to the subject that fail to place it in the context of organic chemistry. Unfortunately, several unifying concepts are covered in detail only in graduate chemistry courses—particularly the utility of molecular orbitals in understanding the ionization process (Chapter 3), the role of competing energy demands on the interplay between various fragmentation processes and the ions that result (Chapter 3), and the concept of "electron pushing" for rationalizing reaction mechanisms (especially Chapters 3 and 7). I have tried to present that material here, written from the perspective of a practicing analytical organic chemist, in a manner that is approachable even for those readers who have not had graduate chemistry courses. The book is thus designed to serve as a textbook for upper-level undergraduate and beginning graduate courses and as a self-help resource book for mass spectrometry practitioners, in laboratories like my own, who desire and need to understand their subject more fully.

I believe that no chemist who uses a mass spectrometer on a regular basis should be ignorant of how it works; without this knowledge, it is difficult to gain an adequate appreciation of the importance of even basic tasks such as instrument calibration. The first chapter thus deals with instrumentation, with an emphasis on a basic understanding of the theory involved. Although many chemists are horrified by mathematical derivations, I hope my own background in mathematics will ease the reader through some of the simpler ones. My intent in including them is to show that at least some of the theoretical underpinning of mass spectrometry is understandable (and thus more easily committed to memory) without advanced mathematical training.

The same approach is taken in the discussion of isotopic abundances (Chapter 2). Instead of presenting a few well-known equations on faith, I have tried to develop at some length the background against which the mathematics can be applied, again with the intent of providing a better understanding of the relationship between isotopic abundances and relative peak sizes.

The number of chapters describing actual types of fragmentation reactions is limited (Chapters 4 to 6) in order to provide a unified approach to the subject, rather than a presentation of disparate pieces of information. Not all reaction types are covered, since I felt it was more important for the reader to fully understand a few fragmentations that have wide application than to try to cover every eventuality. Particular emphasis is placed on α-cleavage, fragmentations that eliminate small unsaturated molecules, and optimal ring sizes for hydrogen rearrangements. I have tried to repeat these concepts in as many contexts as possible throughout Chapters 4 through 8 to emphasize their utility and to facilitate committing them to memory.

Although the stated focus of this book is a *basic* understanding of all mass spectra, many of the examples are derived from my own experience in forensic science. I have found in teaching this subject an additional barrier to understanding—namely that students are not always able to make the leap from simple examples to complex, real-life problems. Many of the compounds that are encountered by chemists in the forensic and health-related sciences superbly exemplify the basic arsenal of mass spectral fragmentations and also add a unique element of interest to the subject matter—especially the spectra of illicit drugs. Chapter 8 focuses on using derivatives of some of these compounds to predict spectra for other compounds having similar structures.

I have purposely shied away from detailed descriptions of advanced topics, especially the variety of new instrumentation available, since they do not substantially alter the basic approach to mass spectral interpretation. More to the point, I felt they would detract from the fundamental nature of the material I am trying to cover. Thus, recent inlet techniques like high-performance liquid chromatography/mass spectrometry (HPLC/MS) and capillary electrophesis/mass spectrometry (CE/MS), and ionization methods such as electrospray, fast atom bombardment, and laser desorption, are given no more than passing mention, and descriptions of high-resolution mass spectrometry and MS/MS are limited.

No matter how proficient I feel I may have become at interpreting spectra, I am still jolted back to reality on a regular basis by compounds that just don't "behave" as I thought they would. Mass spectral interpretation is neither a static nor a shallow subject. Each time I taught this material, and even as I wrote this book, I reached new levels of understanding of even some of the most basic concepts that are presented here. Thus I doubt, for most readers, that the contents of this book will be thoroughly digested in one reading. Rather I would suggest studying it slowly, and even repetitively, with particular emphasis on understanding the problems and on writing mechanisms, applying each concept to the spectra encountered in the specific laboratory situation. The rewards are well worth the effort.

R.M. SMITH

*Madison, Wisconsin*

# ACKNOWLEDGMENTS

There are many people to thank for making this book a reality. Foremost among these are Dave Steingraber, Mike Roberts, and other members of the Wisconsin Department of Justice, Division of Law Enforcement Services, who allowed me to use both time and equipment to develop the materials contained in this book, as well as to write and assemble the manuscript. Without their backing, this book may never have been published. A special thanks goes to my immediate boss, Jerry Geurts, for his support and encouragement.

The contributions of many of my colleagues must be acknowledged also. Any list I make will necessarily be incomplete (there have been so many over the years), but Dave Picard, Ken Kempfert, Mike Haas, Bob Block, Raemarie Szymanski, Marty Koch, Lori Nelsen, Guang Zhang, Casey Collins, Paula Smith, Mike Poquette, and John Nied deserve special mention for providing interesting problem samples that found their way, directly or indirectly, into this book. I particularly want to thank Guang Zhang and Stephanie Ross for reading parts of the manuscript and helping me define the audience I should address.

I owe much to my many students as well, especially those who had the courage to ask "dumb questions." Their probing consistently made me evaluate what I really knew about mass spectral interpretation and forced me to look for better ways to understand and teach it. This book has, in many ways, been shaped by their needs.

Finally, I want to thank Barbara Goldman of John Wiley & Sons—first, for taking the time to read an unsolicited manuscript from an unknown author, and, second, for getting Ken Busch involved in this project. His encouragement, his insightful comments on the overall contents of the book, and particularly his technical editing, criticisms, and contributions to the first chapter were substantial, and are deeply appreciated.

# UNDERSTANDING MASS SPECTRA— A BASIC APPROACH

# CHAPTER 1

# INSTRUMENTATION

## 1.1. INTRODUCTION

The interpretation of mass spectra does not require an understanding of the physics and electronics underlying mass spectrometry. Nevertheless, a basic knowledge of mass spectrometers and the theory behind them is desirable for a number of reasons. First, it gives the user the ability to evaluate instrument performance (perhaps even to troubleshoot problems) and to help maximize the information that can be gained from this powerful and complex analytical tool. Just as importantly, it can also lead to an appreciation of why some ions are seen in the mass spectrum and others are not.

Mass spectrometry (MS) differs from other common forms of organic spectral analysis in that the sample does not absorb radiation from the electromagnetic spectrum (infrared, ultraviolet, radio waves, etc.). Further, in contrast to infrared (IR) or nuclear magnetic resonance (NMR) spectrometry, both of which identify compounds with specificity comparable to that of mass spectrometry, MS is also a destructive method of analysis—that is, the sample cannot be recovered after mass spectral analysis. Fortunately, MS is also highly sensitive, so that the entire sample will be used only when sample size is extremely limited.

During mass spectral analysis, sample molecules first undergo reaction with an ionizing agent, which in the case of electron ionization mass spectrometry (EIMS) is a beam of high-energy electrons. Ionization is followed by fragmentation in which bonds break, and in many instances new bonds form, in ways that are characteristic of the structure of the fragmenting ion. This is why mass spectrometry, especially when coupled with separation techniques such as gas chromatography (GC) or high-performance liquid chromatography (HPLC), is a highly specific way to identify organic compounds.

1

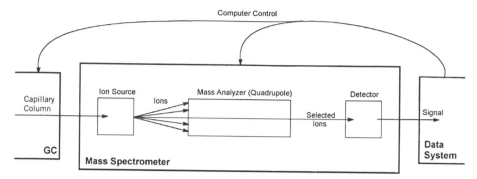

**Figure 1.1.** Block diagram of a GC/MS/computer system.

Figure 1.1 shows a block diagram of a typical computerized gas chromatograph/mass spectrometer (GC/MS). This system consists of three separate units that are coupled together physically and/or electronically. A gas chromatograph separates mixtures and introduces sample molecules in the gas phase into the mass spectrometer; the mass spectrometer ionizes the vaporized sample molecules, then analyzes and detects the resulting ions; and a computer system controls the operation of the GC and the MS, as well as providing data manipulation during and after data collection.

The components of the mass spectrometer that provide ion formation, separation, and detection are contained in an ultraclean metal housing kept at moderately high vacuum ($10^{-5}$ to $10^{-6}$ torr; a few exceptions to this will be mentioned later). High vacuum is necessary for a number of different reasons. First, it ensures that molecules entering the ion source will be in the gas phase and will not condense on the inside surfaces of the spectrometer. Second, the mean free path of ions formed under high vacuum is relatively long, so that collisions with other ions or molecules are extremely rare. Ion focusing and mass analysis requires that the ions formed in the ion source maintain their integrity until they reach the detector. As a result, all of the fragmentation reactions occurring under these conditions are *intra*molecular (involving only the decomposition of individual ions) rather than *inter*molecular (involving the reaction of ions with other ions or neutral molecules or fragments). Finally, high vacuum also protects the metal and oxide surfaces of the ion source, analyzer, and detector from corrosion by air and water vapor, which would seriously compromise the spectrometer's ability to separate and detect ions.

In the system shown in Figure 1.1, compounds are separated by gas chromatography prior to introduction into the mass spectrometer. High sample purity is critical for unambiguous identification by mass spectrometry since the simultaneous presence of several different compounds in the ion source means that ions from all of them are formed and analyzed at the same time—resulting in a composite mass spectrum that may be impossible to interpret. Capillary GC columns maximize the separation power of the chromatograph but demand small amounts of sample. However, because the carrier gas flow through a capillary column is low enough that the carrier gas can be removed by the vacuum system of the mass spectrometer, sample molecules can be introduced into the ion source of the spectrometer directly through the end of the capillary column itself. Helium and hydrogen are good choices as carrier gases for GC/MS work because of their extremely low atomic and molecular weights, which fall below those of all of the ions normally seen in organic mass spectrometry.

In addition to GC, HPLC has become increasingly important as an option for sample separation prior to mass spectral analysis, especially for compounds that are nonvolatile, thermally labile, or for other reasons are not amenable to analysis by GC. More recently, capillary electrophoresis (CE) has been coupled with mass spectrometry to separate and identify inherently ionic molecules such as amino acids, proteins, and even DNA fragments. In contrast to the facile separation of sample and carrier gas in GC/MS, separating sample molecules from HPLC or CE solvents is difficult, so that the development of these techniques as routine methods of compound separation for mass spectrometry has occurred more slowly. The initial ionization processes in both HPLC/MS and CE/MS differ substantially from those in EIMS (Table 1.1) and usually lead to fewer fragment ions in the spectrum.

Other methods of sample introduction must be mentioned briefly. For analysis of a pure volatile liquid, the liquid can be placed in a small, evacuated glass bulb that is connected to the ion source, often with narrow metal or glass tubing, but isolated from the MS vacuum system by a valve. Opening the valve allows the sample to enter directly into the ion source. This method is used for introduction of perfluorotri-*n*-butylamine (PFTBA), the standard used for MS calibration and tuning.

Relatively nonvolatile or thermally unstable compounds, on the other hand, can be placed on the tip of a rod or probe that is inserted directly into the ion source. High vacuum is maintained by a vacuum lock—that is, the rod tip containing the sample is inserted into a chamber that is isolated from the main vacuum system by a valve. This chamber is evacuated using an auxiliary vacuum pump, after which the valve is opened and the probe tip is inserted all the way into the ion source. Gentle heating of the probe tip provides volatilization of the sample and, in ideal cases, even rudimentary fractional distillation of the desired compound. Nonetheless, sample purification prior to introduction by heated probe is desirable. Because of the versatility of capillary GC and HPLC, heatable direct insertion probes have become increasingly rare. Compounds that are entirely nonvolatile (e.g., large biomolecules or ionic compounds) may require special sample introduction and ionization techniques in order to produce ions for mass spectral analysis. A list of ionization methods and their application to various sample types is given in Table 1.1.

## 1.2. ELECTRON IONIZATION SOURCE

Regardless of how they are introduced, sample molecules enter the mass spectrometer through the ion source. Except in the case of the ion trap mass spectrometer (Section 1.3.3), the electron ionization *ion source* is a small, approximately cubical, chamber about 1 cc in volume, in which sample molecules are bombarded with a beam of highly energetic electrons [70 electron volts (eV); 1 eV = 23 kcal]. Since bond strengths in organic compounds range from approximately 10 to 20 eV, this method

**Table 1.1. Molecular Ionization Methods in Mass Spectrometry**

| Type of Ionization | Ionizing Agent | Source Pressure | Uses |
|---|---|---|---|
| Electron ionization (EI) | 70-eV electrons | $10^{-4}$–$10^{-6}$ torr | Extensive fragmentation allows structure determination; GC/MS |
| Chemical ionization (CI) | Gaseous ions | ~1 torr | Molecular weight determination; GC/MS |
| Desorption ionization (DI) | | $10^{-5}$–$10^{-6}$ torr | Molecular weight and structures of higher mass nonvolatile compounds in condensed phase; may use specific type of mass analyzer |
| Fast atom bombardment (FAB) | Energetic Ar or other neutral atoms | | |
| Laser desorption (LDI) and matrix-assisted LDI (MALDI) | Energetic photons | | |
| Spray ionization (SI) | | 1–760 torr | Solutions of high mass compounds; molecular weight and structure (HPLC/MS) |
| Thermospray (TS) | Thermal; gaseous ions | 1–10 torr | |
| Electrospray (ES) | Electric field; ions in solution | Atmospheric to slightly reduced pressure | HPLC/MS and CE/MS |
| Atmospheric pressure chemical ionization (APCI) | Corona discharge; gaseous ions | Atmospheric | HPLC/MS |

of ionization not only strips electrons from molecules, it also makes available to the resulting ions enough energy that substantial fragmentation of the first-formed ion (called the molecular ion) usually occurs. Ion sources from different instrument manufacturers (and sometimes even different models from the same manufacturer) may differ from one another both in appearance and in the names assigned to the component parts, but most have the same basic design. A typical example is shown in Figure 1.2.

After introduction into the ion source, sample molecules are bombarded by an electron beam produced by a filamentous strip of wire (sometimes coiled) made of a tungsten–rhenium alloy. Between the filament and the center of the ion source is a metal plate containing a slit called the electron aperture. This slit limits the size of the electron beam and confines ionization to a small volume within the center of the ion source. Opposite the filament is the collector, a metal plate held at a positive electrical potential ($+V$ in Figure 1.2) that intercepts the electron beam after it has passed through the source. Surrounding the entire ion source in some cases is a collimating magnet, which causes the actual path of the electrons in the beam to become helical, as shown in Figure 1.2. Since sample ionization is very inefficient (fewer than one molecule in a thousand undergoes ionization), this helical trajectory improves the probability that the electrons and molecules will interact.

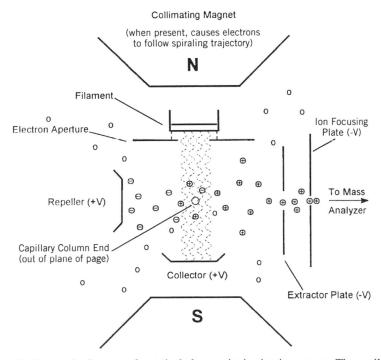

**Figure 1.2.** Schematic diagram of a typical electron ionization ion source. The capillary GC column enters the source from behind the plane of the page. Samples can enter the source from a heated probe or evacuated bulb through openings in front of the plane of the page.

What actually happens during ionization is very complex. It is naïve to view the electron beam as literally smashing into sample molecules and knocking electrons out of orbitals. Instead, ionization is governed by quantum mechanical processes that allow energy transfer to occur without actual "physical contact" between the electrons and the sample molecules. All that is necessary is that an energetic electron approach the electron density field of the molecule closely enough that sufficient energy is transferred to overcome the *ionization potential* of the molecule—the amount of energy needed to remove one electron from one of the bonding or nonbonding orbitals of the molecule. Because of the excess energy present in 70-eV electrons, enough additional energy may be transferred to overcome the second, or even third, ionization potential of the molecule (leading to ions having $+2$ or $+3$ charges) or to produce substantial fragmentation of the initially formed molecular ion. A rudimentary knowledge of the molecular orbitals of the sample molecule allows us to predict the probable site of initial ionization (Section 3.2).

Many different products, not all of which are positive ions, form during ionization. Table 1.2 lists the most important of these. If the sample molecule absorbs only enough energy to raise an electron from the ground state to an excited state, but not enough to cause actual ejection of the electron, an "excited molecule" is formed (product *a* in Table 1.2). Excited molecules can return to their neutral ground state through thermal vibrations or the emission of light, and, since no ions are formed in the process, they are simply pumped away from the ion source by the vacuum system.

Alternatively, an electron may be absorbed by the molecule, forming a negative ion (Table 1.2, product *b*). By reversing the polarity of the repeller, ion focusing plate, and extractor plate in the ion source, and adding a dynode to the detector to convert

**Table 1.2. Types of Ionization Reactions**[a]

| | |
|---|---|
| | *a.* Excited molecule (not detected) |
| | *b.* Negative ion formation (not detected by positive ion EIMS): |
| | $(A–B–C)^-$ |
| | $(A–B)^- + C$ and others |
| | *c.* Ionization: $\mathbf{(A\text{–}B\text{–}C)^+} + 2e^-$ |
| | *d.* Dissociative ionization: |
| $e^- + (A–B–C) \quad \rightarrow$ | $A + \mathbf{(B\text{–}C)^+} + 2e^-$ |
| | $\mathbf{(A\text{–}B)^+} + C + 2e^-$ |
| | $(A–B) + \mathbf{C^+} + 2e^-$ and others |
| | *e.* Dissociative ionization with rearrangement: |
| | $\mathbf{(A\text{–}C)^+} + B + 2e^-$ |
| | $(A–C) + \mathbf{B^+} + 2e^-$ |
| | *f.* Multiple ionization: |
| | $\mathbf{(A\text{–}B\text{–}C)^{2+}} + 3e^-$ |
| | $\mathbf{(A\text{–}B)^+} + \mathbf{C^+} + 3e^-$ and others |

[a]Ions detected by positive ion electron ionization mass spectrometry are shown in boldface.

the negative ions into positive ions (see Section 1.4), a negative ion mass spectrum can be recorded. For most compounds negative ion MS offers few advantages over, and tends to be less sensitive overall than, positive ion MS. There are some specific applications, however, most notably with halogenated compounds. In any case, an electron absorbed by a sample molecule must be of very low energy (approximately 0 to 2 eV), and there are few electrons of this energy in a standard electron ionization source. In this book we will only discuss positive ion products and their fragmentations.

All of the remaining products listed in Table 1.2 are positive ions. The first formed of these ions is that resulting directly from ejection of a single electron from the parent molecule (product *c*). This *molecular ion* is very important since it has the same mass as the molecular weight of the compound. Indeed, mass spectrometry is one of the few analytical tools we have for determining the molecular weight of a compound.

Ion products *d* and *e* (Table 1.2) are formed by unimolecular dissociation of the molecular ion, either by breaking one bond and losing a neutral group ("simple cleavage") or by a process in which some bonds are broken and new bonds are formed ("dissociation with rearrangement"). The equations in Table 1.2 imply that such ions are formed in a concerted process in which ionization, bond making, and bond breaking all occur at about the same instant. However, many, if not most, of these fragmentations occur in a "stepwise" fashion through one or more intermediates, as we shall see in later chapters.

If more than one electron is lost from the parent molecule, ions having charges of $+2$, $+3$, or even $+4$ may be formed (Table 1.2, products *f*). Since mass spectrometry actually measures the mass-to-charge ratio ($m/z$) of an ion, not its mass, an ion having a charge greater than $+1$ is found not at the expected mass ($m$), but rather at $m/2$, $m/3$, or $m/4$, depending on the charge. Further, if $m$ is odd, $m/2$ has a nonintegral value. For example, the doubly charged molecular ion of a compound having a molecular weight of 179 is found at $m/z$ $179/2 = 89.5$. Not all compounds give multiply charged ions; they occur most commonly in the spectra of compounds having extensive aromatic systems or large numbers of heteroatoms (atoms other than carbon and hydrogen). In addition, many mass spectrometers used in organic analytical labs only report $m/z$ values to the nearest integral mass, so that detecting ions at nonintegral masses, even if they occur, is not always possible. Mass spectrometers having higher mass resolution are generally necessary to identify these ions with certainty.

In contrast to the unimolecular ionization processes listed in Table 1.2 for electron ionization mass spectrometry, ionization of sample molecules in chemical ionization mass spectrometry (CIMS) depends on reactions between ions and molecules. In CIMS the sample is ionized by reaction with ions generated within a large excess of a reagent gas such as methane (as $CH_5^+$), isobutane [as $(CH_3)_3C^+$], or ammonia (as $NH_4^+$), at a pressure of about 1 torr. Some reagent gas ions themselves are formed by ion/molecule reactions, viz.,

$$CH_4 + e^- \rightarrow CH_4^{+\cdot} + 2e^-$$

$$CH_4^{+\cdot} + CH_4 \rightarrow CH_5^+ + CH_3\cdot$$

although the reaction

$$(CH_3)_3CH + e^- \rightarrow (CH_3)_3CH^{\ddagger} \rightarrow (CH_3)_3C^+ + H\cdot$$

is unimolecular.

Because sample molecules (at about $10^{-6}$ torr) are so greatly outnumbered by reagent gas molecules in CIMS, the reagent gas is ionized preferentially by the electron beam, and sample molecules become ionized through reaction with reagent gas ions, rather than by the electron beam itself. Most reagent gas ions are strong acids, and form *protonated molecules* (previously called pseudomolecular ions) that have a mass 1 dalton (1 dalton = 1 atomic mass unit) greater than that of the molecular weight of the original compound, representing the transfer of a proton to the neutral sample molecule acting as a base[1]:

$$M + CH_5^+ \rightarrow MH^+ + CH_4$$

The protonation reactions impart much less energy to sample molecules than do interactions with 70-eV electrons, so that this process (often called "soft ionization") forms sample ions that have little excess internal energy and therefore fragment to a lesser degree than do ions formed by EIMS. As a result, CIMS is useful for determining the molecular weight of compounds that produce only weak molecular ions by EIMS. The protonated molecules formed during CIMS still dissociate only by *intra*molecular fragmentation processes that are analogous to those of the molecular ions produced in EIMS.

The complex mixture of ionic and neutral products formed by any of these ionization methods must be separated so that positive ion products travel in the direction of the mass analyzer, leaving behind negative ion and neutral products. Neutral products are removed by the vacuum system, since the electric and magnetic fields present in the ion source have no effect on their motion. Positive and negative ions, on the other hand, can be separated by appropriately placed charged surfaces in the ion source (Figure 1.2). To accomplish this, the repeller is kept at a positive potential ($+V$) both to attract and neutralize negative ion products, as well as to repel positive ions. Conversely, the extractor plate and ion focusing plate are both kept at a negative electrical potential ($-V$) to attract and accelerate the positive ions toward the mass analyzer. Slits cut into the extractor and ion focusing plates allow passage of the positive ions and focus the ion beam as it approaches the analyzer.

When the filament is on and sample molecules are flowing into the ion source, many reactive species are produced. Indeed the intensity of the electron beam itself is sufficient to corrode metal surfaces in the ion source that are directly in its path—those on the electron aperture and collector. In addition, many ion products become electrically neutralized and polymerize on the surfaces of the repeller, extractor plate, and ion focusing plate. Over time, the sensitivity of the instrument declines as these

[1]Some reagent gas ions may react with sample molecules by addition, rather than by proton donation. It is not unusual to observe small, but significant, ions at masses greater than that expected for the protonated molecule, corresponding to the addition of one or more reagent gas ions to the sample molecule.

surfaces become less able to maintain the potentials necessary for optimal ejection and focusing of positive ions from the source. Mechanical and chemical cleaning of the metal surfaces in the source is needed to restore sensitivity. Keeping the filament off when high concentrations of sample are present in the ion source (especially while solvents are eluting during a GC run) allows the source to remain usable for several months without cleaning. Chemical ionization mass spectrometry, which depends on the presence of high ion concentrations in the source, leads to the deterioration of ion source performance much more rapidly than does electron ionization under normal circumstances.

## 1.3. MASS ANALYZERS

### 1.3.1. Introduction—Magnetic Sector Analyzer

The mixture of molecular and fragment ions formed in the ion source contain information that would be lost were these ions not separated and identified in some meaningful way. In particular, we would like to measure the individual mass of each of the ions present. To do this, the mass spectrometer must be able to vary the mass being detected by changing some physical or electronic property within the instrument—that is, a mass analyzer must use the fact that, in certain environments, charged particles have motions that are directly related to their mass-to-charge ratio ($m/z$; in early literature, $m/e$). Although the quadrupole mass filter has become the most widespread mass analyzer used in GC/MS and HPLC/MS, other types of mass analyzers are currently being used both for these and other applications. These include the magnetic sector and time-of-flight analyzers, both of which played prominent historical roles in the development of mass spectrometry as an analytical tool, and the ion trap, a more recent addition that is actually a variation of the quadrupole mass filter. The magnetic sector analyzer is a convenient example with which to begin, partly for historical reasons, but also for the relative ease with which a mathematical connection can be made between physical motions of ions and their mass-to-charge ratios.

A generalized schematic of a magnetic sector mass spectrometer is shown in Figure 1.3. In this instrument, ions formed in the ion source are accelerated toward the magnet by an accelerating potential $V$. Once the ions come under the influence of the magnetic field $B$, whose lines of force are perpendicular to the plane of the drawing, the ion paths curve along an arc of a circle whose radius is $r$. The mass spectrometric equation for this instrument can be derived from fundamental physics relationships using these variables.

In a magnetic field, an ion with mass $m$ will experience a centripetal force (one pulling the ion toward the center of the circle) equal to $Bzv$, where $B$ is the strength of the magnetic field, $z$ ($= ne$; the number of charges times the charge on one electron) is the charge on the ion, and $v$ is the velocity of the ion. At the same time, any particle moving on a circle having a radius $r$ experiences a centrifugal force (one pushing it away from the center of the circle) equal to $mv^2/r$. When these two forces are equal, the ion travels on the circle and

$$Bzv = mv^2/r \quad \text{or} \quad m/z = Br/v \qquad (1.1)$$

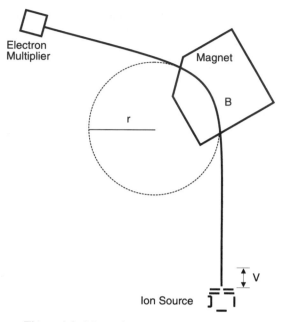

**Figure 1.3.** Magnetic sector mass spectrometer.

We could measure $v$, the velocity, to determine mass, but this is experimentally diffi-cult. Thus, some other expression for this variable must be found that can be substi-tuted into Eq. (1.1). To do this, note that the kinetic energy of the moving ion, once it reaches the magnet, is $\frac{1}{2}mv^2$, which is equal to the potential energy ($= zV$) the ion had when it left the ion source. In other words,

$$\tfrac{1}{2}mv^2 = zV \quad \text{or} \quad v^2 = 2zV/m$$

and

$$v = \sqrt{\frac{2zV}{m}} \tag{1.2}$$

Substituting this expression for $v$ back into Eq. (1.1), we obtain

$$\frac{m}{z} = \frac{Br}{\sqrt{2zV/m}}$$

or, squaring both sides,

$$m/z = B^2r^2/2V \tag{1.3}$$

Since the strength of the magnetic field, the radius of the circle, and the accelerating potential are all measurable quantities, we now have a method for determining the $m/z$ of the ion.

In order to measure $m/z$ and obtain a mass spectrum, two of the variables in the right-hand side of Eq. (1.3) must be held constant. There are several ways of doing this in a magnetic sector instrument. In one geomtery, both the magnetic field and the accelerating potential are held constant. Then, ions having different $m/z$ values pass through the magnetic field along paths with different values for the radius $r$. Instead of an electron multiplier detector (which only detects ions traveling along a single path of fixed radius $r$; Section 1.4), a high-resolution photographic plate is placed perpendicular to the paths of the ions after they have passed through the magnet. As a result, the position at which an ion collides with the photographic plate is related to its $m/z$ value, and the mass spectrum is obtained by developing the photographic plate. Despite problems in assigning $m/z$ values to each peak in the spectrum ($m/z$ varies linearly with $r^2$, not $r$) and determining relative ion intensities from the densities of the spots on the developed photographic plate, this method has proven to be extremely useful for many years and provides a high degree of mass resolution. This early form of array detection has been revisited recently with the development of photodiodes as detectors for mass spectrometry.

Another common arrangement of the magnetic sector mass spectrometer holds the accelerating potential constant, while requiring a constant value for $r$ by using a narrow slit followed by an electron multiplier detector. By scanning the magnetic field $B$, ions of sequential $m/z$ values attain the appropriate value for $r$, and then pass through the slit to the detector. At any given value of $B$, all other ions having different trajectories (i.e., different $r$) collide with the inner surfaces of the spectrometer. As with the arrangement above, however, $m/z$ varies linearly with $B^2$, making assignment of $m/z$ values somewhat difficult. Even more importantly, electromagnets do not return immediately to their original state after having had their field strength changed because of a phenomenon called hysteresis. Hence each spectral scan takes several seconds, with a substantial portion of the scan spent waiting for the magnet to equilibrate. The advent of combined GC/MS, which demands rapid scan rates (1 s/scan or less) in order to maintain chromatographic resolution, forced improvements in magnet technology to increase scan speeds and limit the effects of hysteresis. Today, magnets scan at a speed fast enough for all but the most demanding chromatographic separations.

In the late 1970s, an attempt was made to allow magnetic sector mass spectrometers to compete with quadrupoles for the GC/MS market by holding $B$ constant while scanning through the accelerating potential $V$, which can be changed rapidly (about 1 s/scan). By this time computer technology also had become sophisticated enough to handle instrument tuning and data collection (minimizing the effect of the inverse relationship between $m/z$ and $V$). Nonetheless, scanning below $m/z$ 43 generated accelerating potentials in excess of 12,500 V in the ion source, sometimes causing arcing and also requiring retuning several times during the course of a day.

### 1.3.2. Quadrupole Mass Analyzer

Quadrupole mass spectrometers have taken over the GC/MS market over the past 25 years. As is seen in Figure 1.4, the quadrupole mass filter consists of four metal rods held in strict alignment with one another—indeed it is crucial that they remain parallel to and at a fixed distance from one another. Opposing rods are connected in pairs to both radio frequency (*RF*) and direct current (dc) generators, bathing ions in a combined electric and radio frequency field during their passage through the rods. The output of the *RF* generator is simply energy in the radio frequency part of the electromagnetic spectrum, and can be pictured as a sinusoidal wave having a zero-to-peak amplitude of *U* and a frequency ω (the number of wave crests per second). The output of the dc generator, on the other hand, is simply two different voltages, + *V* and - *V*. The amplitude of the *RF* output can be changed even while the frequency is held constant.

In contrast to the equation of motion derived in the previous section for ions passing through a magnet, those for quadrupole mass analyzers are very complex. In fact, their complete derivation, which involves solution of the differential equations that describe the motions of ions in combined electromagnetic and electric fields, is well beyond the scope of this book. This derivation generates two combination variables *a* and *q*:

$$a = \frac{8zU}{mr^2\omega^2} \tag{1.4}$$

$$q = \frac{4zV}{mr^2\omega^2} \tag{1.5}$$

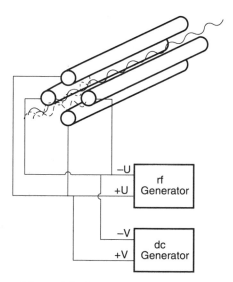

**Figure 1.4.** Quadrupole mass analyzer.

where $U$ = zero-to-peak voltage of the applied radio frequency field
   $V$ = applied dc voltage
   $r$ = radius of the circle tangent to the inner surfaces of the quadrupole rods,
   $\omega$ = applied radio frequency
   $m/z$ = mass-to-charge ratio of the ion
It is important to recognize that $a$ and $q$ have no physical meaning in the instrument.

In these equations $r$ and $\omega$ are held constant (the former because of the physical design of the instrument and the latter by choice), but some further constraint must be made on the system in order to determine $m/z$ as a function of a single variable. In this case $a$ and $q$ are chosen to be proportional to one another, so that

$$\frac{a}{q} = \frac{2U}{V} = \text{constant}$$

Then, since $a$, $q$, $U$, and $V$ are now all interrelated, $m/z$ varies with either $U$ or $V$. This is particularly convenient since $U$ and $V$ are both easily controllable by electronic circuitry.

The relationship between all of these variables and the determination of $m/z$ can best be understood in terms of a plot of $a$ vs. $q$ (Figure 1.5). Actually this is only a small portion of the entire plot for all values of $a$ and $q$; for reasons involving instrument design, only the part of the plot near the origin is of interest. Only the shaded area of the graph contains values for both $a$ and $q$ that define stable ion motion along the $z$ axis (the axis parallel to the four rods), thereby allowing ions to pass through the rods to the detector. For other values of $a$ and $q$, ions wander far enough from the $z$ axis that they collide with the rods.

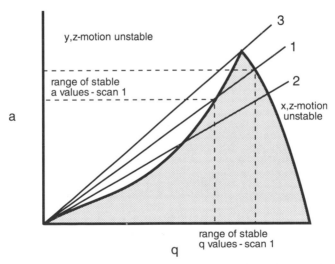

**Figure 1.5.** Plot of $a$ vs. $q$ for quadrupole mass spectrometry. Choosing scan line 3 fixes values for $a$ and $q$ so that $m/z$ varies linearly with either of the voltages $U$ or $V$.

The lines marked 1, 2, and 3 in Figure 1.5 are called scan lines. All scan lines in this plot are straight lines passing through the origin and having a slope $a/q$ that is fixed instrumentally by choosing $2U/V$ ($= a/q$) as a constant. It is important to note that neither $a$ nor $q$ is a function of time—that is, a spectrum is not collected by starting at the left end of a scan line and following it through higher values of $a$ and $q$. It is equally important to recognize that the values of $a$ and $q$ are not determined when the instrument is "tuned" on a daily basis. These values are set indirectly on installation when parameters are found that produce observable ions with reasonable mass resolution with the instrument.

The actual scan line used experimentally is determined by the desired mass resolution. In mass spectrometry, *resolution* is defined as $M/\Delta M$, where $M$ is the mass of the ion and $\Delta M$ is the smallest increment of mass that can be distinguished by the analyzer. For quadrupole analyzers $\Delta M \sim 1$ dalton over the entire mass range (often improperly referred to as "unit resolution"). Nonquadrupole instruments offering $\Delta M$ values of 0.001 or less are available but are also substantially more expensive (see Section 1.3.4). The utility of high-resolution mass spectrometry will become apparent in later chapters. Analyzers with $\Delta M > 1$ have only limited utility since individual masses are then no longer distinguishable.

Choosing scan line 1 in Figure 1.5 defines ranges of values for both $a$ and $q$ that allow stable ion motion through the quadrupole at any instant. Since $a$ and $q$ are both functions of $m/z$, all ions within the range of $m/z$ values defined by the "allowable" values for $a$ and $q$ traverse the rods at the same time; if the range is greater than 1 dalton, we cannot determine the masses of the individual ions. Scan line 2 is worse since it allows the passage of even more ions at the same time. The ideal choice, therefore, is scan line 3, which passes through the apex or cusp of the area defining stable motion along the $z$ axis. Solving Eqs. (1.4) and (1.5) for $m/z$, we have

$$\frac{m}{z} = \frac{8U}{ar^2\omega^2}$$

and

$$\frac{m}{z} = \frac{4V}{qr^2\omega^2}$$

By passing through a unique point of $z$-stable motion on the graph, scan line 3 fixes unique values for both $a$ and $q$ and thereby renders them constant. Since $r$ and $\omega$ were fixed when the instrument was set up, $m/z$ then becomes dependent upon only one variable—either $U$ or $V$, since $U$ and $V$ are chosen to be dependent upon one another. Thus,

$$m/z = kU \quad \text{or} \quad k'V \tag{1.6}$$

where $k$ and $k'$ are constants.

Equation (1.6) is particularly convenient because $m/z$ varies linearly with voltage, which is not only easy to control instrumentally but also can be varied rapidly over a scan with little or no lag time between scans. In the Hewlett-Packard Mass Selective Detector, the dc generator is scanned from approximately 0 to 200 V and the RF generator from about 0 to 1200 V, producing a range of $m/z$ values from 0 up to about 800 daltons. With GC/MS, it is desirable not to scan below $m/z$ 10 because of the high concentration of helium ($m/z$ 4) or hydrogen ($m/z$ 2) in the ion source. Many users routinely do not scan below $m/z$ 35 to avoid air and water background in their spectra. The mass ranges of most quadrupole instruments extend up to $m/z$ 800 to 1000, although instruments with mass ranges up to $m/z$ 2000 are available.

Experimentally, scan line 3 cannot be followed precisely (i.e., so that the scan line passes exactly through the cusp of the $a$ vs. $q$ plot). If that were possible, the instrument would have extremely high resolution but allow so few ions through the analyzer at any time that sensitivity would become a serious problem. Some mass resolution is thus sacrificed in order to gain usable sensitivity.

One advantage of having $m/z$ dependent upon an easily controllable variable such as $U$ or $V$ is that the dc and RF generators can be programmed to produce only discrete values for $U$ or $V$, thereby allowing only ions of specific $m/z$ values to traverse the analyzer and rejecting all others. This process, called *selected ion monitoring (SIM)*, offers enhanced sensitivity for detecting extremely low concentrations of compounds in samples. This occurs because, when the instrument repeatedly scans full spectra (e.g., from $m/z$ 35 to 400), most of the time is spent collecting information about $m/z$ values for which few or no ions are formed by the compound in question. Thus, despite high concentrations in the ion source of fragment ions that are abundant in the spectrum, these ions traverse the analyzer for only very brief periods of time and, as a consequence, only a small fraction of them ever reach the detector. Selected ion monitoring allows only the most intense (and, for identification purposes, the most characteristic) ions to pass through the analyzer more often and increases the amount of time (called the "dwell time") that the analyzer remains at a given value of $U$ or $V$ (and thus $m/z$). This substantially increases the fraction of these ions that reach the detector. Thus while subnanogram ($10^{-10}$ to $10^{-9}$ g) quantities of compounds normally can be detected using full-scan quadrupole mass spectra, SIM can sometimes lower the limit of detection into the picogram ($10^{-12}$ to $10^{-11}$ g) range.

### 1.3.3. Ion Trap Analyzer

The ion trap, despite its appearance (shown in cross section in Figure 1.6), is in fact a variation of the quadrupole. It is a relatively recent addition to mass spectrometric instrumentation, with capabilities the quadrupole cannot offer. Although the dome-shaped end caps and toroidal (roughly doughnut-shaped) ring electrode provide radio frequency and/or electric fields, ions do not "pass through" the analyzer as they do in the magnetic sector and quadrupole instruments. Instead the central area of the ion trap serves as both ion source and analyzer. Ionization is provided by electrons emitted from a filament imbedded in the upper end cap and focused through an aperture in the end-cap surface. There is a similar opening in the lower end cap to allow ions to reach the detector.

For ordinary GC/MS, the dc generator is not used and the ions are subjected only to a radio frequency field. The equations of motion for the ion trap are similar to those

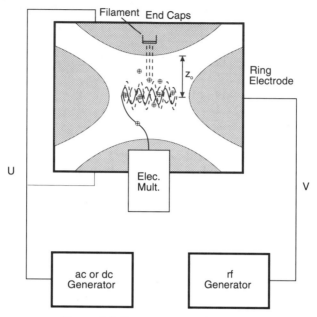

**Figure 1.6.** Ion trap mass spectrometer.

for the quadrupole, except that they relate only to motion in the $z$ direction (which, in this case, is perpendicular to the surfaces of the end caps):

$$a_z = \frac{-8zU}{mr_0^2\omega^2}$$

$$q_z = \frac{4zV}{mr_0^2\omega^2} \tag{1.7}$$

where $U$ = dc (or ac) voltage applied to the end caps

$\quad$ $V$ = RF voltage applied to the ring electrode

$\quad$ $z_o$ = distance from the closest approach of the end cap to the center of the ion trap

$\quad$ $r_o = (2z_o)^{\frac{1}{2}}$

Note that the symbols for the dc and RF voltages are reversed in these equations compared to those for the quadrupole.

Consider the case in which $U = 0$ (i.e., the dc or ac generator is turned off, the configuration normally encountered in routine GC/MS use). Then $a_z = 0$ and $q_z$ is directly proportional to $V$ at a given value of $m/z$. The $a$ vs. $q$ plot for the ion trap is virtually the same as for a quadrupole (Figure 1.7a), but since $a_z = 0$, only points along the $q_z$ axis need be considered. Thus an ion having a particular $m/z$ value will have stable motion in the ion trap below some appropriate value of $q_z$ (and thus of $V$). At very low values of $V$, $q_z$ will fall within the stable motion region for all but the small-

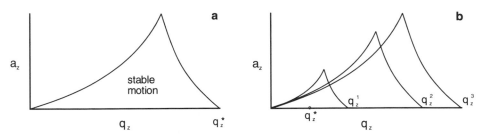

**Figure 1.7.** (*a*) Plot of $a_z$ versus $q_z$ for the ion trap analyzer. For routine GC/MS use, $a_z = 0$; (*b*) $a_z$ vs. $q_z$ plots for different values of $q_z$.

est *m/z* values (Figure 1.7). Under these conditions most of the ions are thus "trapped" into toroidal motions in the center of the analyzer. At high values of $V$, only large values of *m/z* will have stable motions.

For a given value of $q_z$ (arbitrarily designated as $q_z^*$), *m/z* is proportional to $V$ [specifically, $m/z = (4/q_z^* r_0^2 \omega^2)V$]. If we start scanning from low values of $V$, progressively higher and higher *m/z* values reach their limits of stability and "peel off" from the "trapped" ion cloud toward the detector. A full-scan sequence for the ion trap is thus to (a) briefly turn on the filament and form all ions, (b) trap ions using a very low value for $V$, and (c) destabilize ions sequentially by increasing $V$. A relatively high pressure of inert gas (usually $10^{-2}$ to $10^{-3}$ torr of He) is used to dampen the motions of the ions so that they will remain within the trapping field.

The ion trap is not, however, limited to routine GC/MS use. For example, the mass range of this instrument can be extended to *m/z* 70,000 by a process called resonance ejection, which uses an ac, rather than a dc, generator attached to the end caps. Although in normal GC/MS use the ion trap provides mass resolution comparable to that of the quadrupole, the ion trap can be engineered to produce high-resolution mass spectra and to record MS/MS spectra (Section 1.3.4). A detailed discussion of these, and other uses for the ion trap, is beyond the scope of this book.

### 1.3.4. Other Analyzers

***1.3.4.1. Time-of-Flight Mass Analyzer.*** Mass analysis with a time-of-flight (TOF) mass analyzer is based on the simple principle that ions that are given the same kinetic energy will have velocities proportional to their masses. In the ion source of a TOF instrument, ions of all masses are formed almost simultaneously using a very brief burst of energy and then are accelerated out of the ion source by an accelerating potential $V$. The potential energy given to each ion, then, is $eV$, where $e$ is the number of charges on the ion. In the flight tube of the spectrometer, all of this energy appears as kinetic energy in the moving ion ($=\frac{1}{2}mv^2$, where $m$ is the mass of the ion and $v$ is its velocity). In other words, the more massive the ion, the slower it travels. The output of the detector is plotted as a function of time, and this time is converted to mass-to-charge values by the data system. Ions that differ in their flight times by as little as 1 ns can be recorded. After the slowest-moving (highest-mass) ion is detected, the mass analyzer is ready to accept another set of ions accelerated out of the ion source of the spectrometer.

Since the time between ion pulses in the source can be as much as several milliseconds, very high mass ions can be analyzed with the TOF mass spectrometer. Further, the TOF mass analyzer collects *all* of the ions created in each ionization pulse, and thus can be a very sensitive mass analyzer. A few GC/MS instruments have been based on TOF mass analyzers, but most TOF instruments have been built in conjunction with pulsed laser ionization sources. Here the high mass range and sensitivity of the TOF analyzer are useful, and the relatively low mass resolution (caused by the fact that ions are formed at slightly different times and locations in the ion source) is not a problem.

### 1.3.4.2. Fourier-Transform Mass Spectrometers.

Fourier-transform mass spectrometers (FTMS) are significantly more expensive than either quadrupole or sector instruments. In FTMS, ions are created and sent into a cubic, metal cell held in a static magnetic field. For positive ions, small positive potentials on all of the walls of the cell keep the ions near the middle of the cell, where they rotate in circular ion orbits having radii and frequencies that are proportional to their masses. A radio frequency signal is applied to all of the ions in the cell to bring them into phase with other ions of the same mass. The coherent rotating ions induce an oscillating current in the walls of the cube as the ions approach and then recede from the metal plates. For each ion, the frequency of that signal is proportional to the radius of the orbit and the mass of the ion.

All of the frequencies of the trapped ions are recorded together. This combined signal then is subjected to Fourier transformation, a mathematical procedure that reveals the individual masses of the ions by approximating the original signal with a series of sine and cosine functions. If the signal is monitored long enough, the frequencies can be determined very accurately, and exact masses can be calculated. However, keeping the ions in the cell for a long time requires very low pressures (about $10^{-9}$ torr), which is difficult to maintain with a constant flow of carrier gas from a gas chromatograph or a solvent stream from a liquid chromatograph. For this reason, FTMS instruments are not routinely used with GC or HPLC.

### 1.3.4.3. Mass Spectrometry/Mass Spectrometry (MS/MS).

Experiments grouped under the general term MS/MS (or tandem mass spectrometry, signifying two independent stages of mass analysis) help to establish the relationships between ions in a mass spectrum—in particular, what ions are formed when the molecular ion or other ion in the spectrum undergoes fragmentation. This information is extremely helpful in determining overall fragmentation pathways for the molecule being studied. Also MS/MS is widely used to identify specific compounds that are present in complex mixtures.

A typical analyzer arrangement for doing MS/MS uses a linear arrangement of three quadrupoles between the source of ions and the detector (Figure 1.8a). The first and the third quadrupoles act as independent mass analyzers, while the second (middle) quadrupole acts as a collisional activation chamber through which ions from the

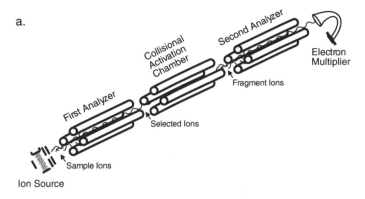

a.

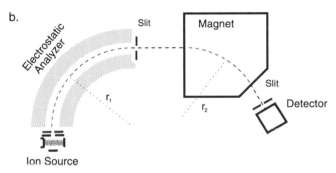

b.

**Figure 1.8.** (*a*) Triple-quadrupole MS/MS in the mode for collecting product ion spectra. Voltages to the first analyzer are fixed to allow transmission of only a single *m/z* value. Collisions of these ions with an inert gas in the middle quadrupole causes further fragmentation, the products of which are analyzed by scanning the second analyzer. (*b*) Double-focusing, high-resolution mass spectrometer of the Nier–Johnson geometry. The electrostatic analyzer focuses the kinetic energy of the ion beam, which allows the magnetic analyzer to identify mass-to-charge ratios over a very narrow range of values.

first quadrupole must pass before they enter the final quadrupole. In a product ion MS/MS spectrum, the first quadrupole ("first analyzer" in Figure 1.8*a*), instead of scanning, is set to pass ions of only one selected *m/z* value (as in SIM, Section 1.3.2) into the collisional activation chamber. The ion so selected is one normally observed in the mass spectrum of the compound—it could be the molecular ion, or one of the fragment ions in the spectrum. In either case, this mass-selected ion is known as the parent ion.

The middle quadrupole is filled with an inert target gas to a pressure that exceeds the normal pressure found in either of the mass-analyzing quadrupoles. Thus when the parent ion passes into the middle quadrupole, a collision between the parent ion and a target gas molecule becomes likely. In such a process, part of the kinetic energy of the parent ion is converted into internal energy, and this internal energy can then cause dissociation of the parent ion. The middle quadrupole is also operated only with RF applied to the rods, that is, with no dc component. As such, it acts as an ion focusing device, conducting both parent ions and the fragment ions formed from them into the final (third) quadrupole.

The last quadrupole ("second analyzer" in Figure 1.8a) is scanned from low mass to a mass just above that of the parent ion, so that the detector records a mass spectrum in which all of the ions must have originated in the parent ion itself. This mass spectrum is called a product ion MS/MS spectrum and provides a direct connection between a parent ion and its fragments.

Production MS/MS spectra can be recorded using mass spectrometers constructed with magnetic and electric sectors, as well as with ion trap and FTMS instruments. In the case of the ion trap and FTMS, ion selection, collisional activation, and product ion analysis can be performed sequentially within the same chamber. Although the operational details differ from instrument to instrument, the basic description and premise of the product ion MS/MS spectrum remain the same.

### 1.3.4.4. High-Resolution Mass Analysis.

We alluded briefly (Section 1.3.2) to mass spectrometers that provide very accurate mass measurement ($\Delta M \ll 1$). In Section 2.1.2, we will discuss how exact mass measurement can be used to distinguish between ion empirical formulas having the same nominal mass ($C_8H_5O_3^+$ vs. $C_9H_9O_2^+$ or $C_{11}H_{17}^+$, e.g., all of which have a nominal mass of 149). Theoretically, several different types of instrumental arrangements can achieve high-resolution mass analysis—FTMS (Section 1.3.4.2) and a specially designed ion trap (Section 1.3.3) have already been mentioned. Higher mass resolution is possible even with a quadrupole analyzer, but as described in Section 1.3.2, the number of ions passing through the analyzer under such conditions drops to an unacceptably low value.

More typically, high-resolution mass spectrometry has been carried out using sector instruments such as the Nier–Johnson double-focusing instrument shown in Figure 1.8b, in which the electric and magnetic sectors work in concert to correct for aberrations in ion optics, and the mass resolution is ultimately defined by the width of the slits along the ion path. Wider slits give lower mass resolution but higher ion throughput, whereas narrower slits provide higher mass resolution, but a decreased ion transmission. Nonetheless, ion transmission through such double focusing sector instruments is usually higher than through quadrupole mass analyzers, and better than unit mass resolution is usually achieved.

### 1.3.5. Concentration Dependence of Ion Intensities

A small but significant amount of time is required for the analyzer to scan through a mass spectrum. Thus if the concentration of the sample changes during the scan, the relative intensities of ions over the entire spectrum will not be reproducible from one spectrum to the next. For GC/MS, particularly with the very narrow GC peaks produced by capillary columns, sample concentrations change rapidly with time, and spectra obtained from different points on the GC peak may look different from one another, despite being spectra of the same compound.

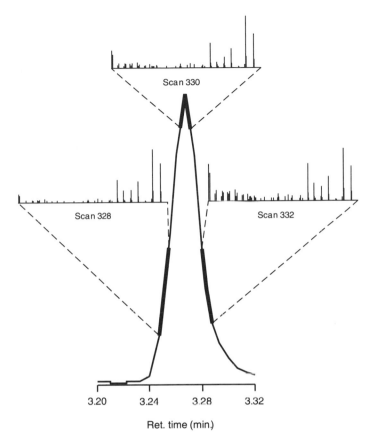

**Figure 1.9.** Variations in mass spectral peak intensities due to concentration differences over a narrow GC peak.

An example of this phenomenon is illustrated in Figure 1.9, which shows a narrow GC peak obtained during an analysis for $\Delta^9$-tetrahydrocannabinol (THC) in marijuana. Notice that over 95% of the sample elutes within 0.06 min = 3.6 s. During this time the mass spectrometer scans through the mass range from $m/z$ 35 to 400 eight times, a scan rate of 3.6/8 = 0.45 s/scan. Nevertheless, from the heavy shading on the front and back sides of the peak, we see that during scans 328 and 332 the concentrations of THC change by roughly a factor of 2 from the beginning to the end of the scan. The result is that the spectra of THC obtained in scans 328 and 332 differ from one another—in scan 328 the high mass ions are intense compared to low mass ions, whereas in scan 332 the low mass ions are nearly as prominent as the high mass ions. Scan 330, obtained at the top of the peak, shows an intermediate situation.

The spectra shown in Figure 1.9 are obtained if the spectrometer is scanned from low mass to high mass (for quadrupoles, from small to larger values of $U$ and $V$; for ion traps, from small to larger values of $V$). Under those conditions the low mass ions in scan 328 reach the detector when the concentration of THC is relatively low, while the high mass ions are detected when sample concentration is higher. In scan 332 the reverse is true. However, some quadrupole instruments (most notably Hewlett-Packard's Mass Selective Detector) scan from high to low mass (high $V$ to low $V$; this cannot be done with the ion trap since, at high values of $V$, none of the ions are "trapped"). In that case, you can convince yourself that the spectra obtained from scans 328 and 332 will be interchanged.

For GC peaks resulting from only a single component, the fact that the spectra change over the course of the peak is not a significant problem since the single scan over the crest of the peak, or a composite spectrum derived from averaging the spectra over the entire peak, will be representative of a "normal" spectrum for that compound. When two or more compounds coelute, however, it may become necessary to use spectra along the sides of the peaks in order to minimize the presence of the spectra of the coeluting substances in that of the desired component.

In theory, the data system should be able to subtract out the contributions from unwanted background spectra, but choosing a good "background" spectrum for subtraction when peaks overlap often proves problematic. In particular, if the desired component elutes first, the concentration of the second component will still be increasing when that of the first has reached its maximum (Figure 1.10). The logical choice for a background spectrum, then, would be the point on the backside of the second peak where the concentration of the second component is approximately equal to that at the point where the spectrum of the first component was taken. However, the two spectra of the second component that are to be subtracted from one another *are not the same* because the concentration of this compound is changing in opposite directions during the two scans. Thus some residual peaks from the spectrum of the second component will remain in the spectrum of the first, despite background subtraction. The only way to eliminate peaks due to the second component may be to choose a spectrum of the first component on the front side on the first peak—but in that case the relative abundances over the entire spectrum may not be representative of "standard" spectra for that compound.

## 1.4. ELECTRON MULTIPLIER DETECTOR

Since ionization in electron ionization mass spectrometry is so inefficient (Section 1.2), the absolute number of ions reaching the detector is small, and some means of amplifying the signal is needed. Consider the situation for a 100-ng sample of a compound having a molecular weight of 200—a large amount of sample for capillary GC/MS work. This sample contains approximately $3 \times 10^{13}$ molecules. Ideally, after injection onto a capillary GC column, a large percentage of these molecules will reach the ion source, although not all at the same instant. A spectral scan at the top of the GC peak may sample only 10 to 20% of the eluting material, lowering by a factor of 5 to 10 the amount available for ionization. If ionization efficiency is as good as 1 molecule in $10^3$ (which is optimistic), then only about $10^9$ molecules are ionized at maximum sample concentration. Further, most of these ions never make it to the detector since the analyzer only allows passage of one $m/z$ value at a time. If the in-

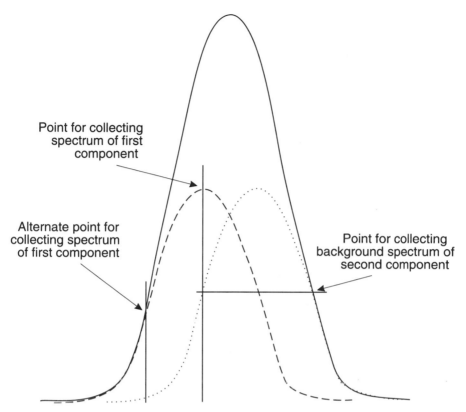

Point for collecting
spectrum of first
component

Alternate point for
collecting spectrum
of first component

Point for collecting
background spectrum of
second component

**Figure 1.10.** Coeluting GC peaks present a special problem for obtaining "acceptable" spectra of the individual components. Background subtraction may not completely rid the desired spectrum of contributions due to the coeluting compound. Getting a "clean" spectrum may be limited to points on the back or front sides of the peak.

strument scans over a range of 300 daltons, then ions having any given $m/z$ value pass through the analyzer only $\frac{1}{300}$ of the time. This leaves a maximum concentration of $10^{13} \times \frac{1}{10} \times (\frac{1}{10})^3 \times \frac{1}{300} \approx 10^6 - 10^7$ of the most abundant ions striking the detector for that scan. While this may seem like a large number of ions, remember that $10^7$ molecules is a meager $10^{-16}$ moles! For a 100-pg sample, only $10^4$ ions will be observed. Since each ion carries a charge of only $1.6 \times 10^{-19}$ coulomb, the current generated even under the best conditions will be very weak.

The electron multiplier detector restores this lost sensitivity by exploiting the ability of certain surfaces to expel more than one electron when an ion collides with it. When the positive ions formed in the ion source and sorted by the analyzer strike the multiplier surface, electrons are ejected from the surface. The conductive surface of the electron multiplier contains a copper–beryllium alloy or a lead-doped glass (about 10 to 20% lead oxide). This surface is somewhat sensitive to air and is especially sensitive to high concentrations of ions, so that venting the vacuum system when the mul-

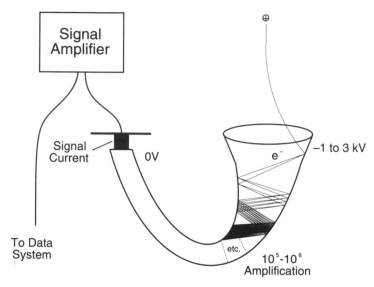

**Figure 1.11.** Continuous dynode electron multiplier detector.

tiplier is on, or (much worse) leaving the filament on while the solvent is eluting during a GC run, can seriously damage the multiplier. Figure 1.11 shows a diagram of a continuous dynode electron multiplier in which the entire surface of the multiplier is physically and electrically continuous. Other, particularly older, types of electron multipliers may have discrete dynodes or stages that are physically distinct but electrically connected to one another.

The surface near the top of the electron multiplier is kept at a highly negative potential (usually $-1.2$ to $-3$ kV). The bottom of the multiplier, on the other hand, is referenced to ground (0 V) so that, to the ejected electrons, the remainder of the multiplier after the top surface looks more positive, and they are attracted further into the multiplier. As each ion or electron collides with the multiplier surface, an average of two electrons are ejected from the surface. The actual number of electrons ejected depends on the gain of the multiplier, which is roughly a function of the total potential difference from "top" to "bottom" of the multiplier surface. The gain is often adjusted during instrument tune-up so that the same amount of a standard sample (usually PFTBA, Section 1.5) will produce approximately the same signal intensity. Each electron ejected by the second collision also causes ejection of two electrons, and this process continues down to the bottom or last dynode of the multiplier. The total signal magnification is thus approximately $2^n$, where $n$ is the total number of collisions with the multiplier surface. Most multipliers provide about a $10^5$- to $10^6$-fold increase in signal, which is comparable to about 18 to 20 collisions. Electrons generated in the last collision with the multiplier surface constitute the signal current output of the multiplier. This current is sent to an external electronic signal amplification circuit and finally to the data system.

## 1.5. DATA SYSTEM

### 1.5.1. Instrument Tuning and Calibration

A mass spectrometer will produce no meaningful information if the analyzer is not set up upon installation to satisfy the equations of motion given in Section 1.3. Beyond this, each mass spectrometer also must be fine-tuned to adjust for the slight imperfections introduced during the manufacture of such a complex unit, as well as for the variations in the ion source, analyzer, and electron multiplier surfaces that occur during routine use. Thus, mass spectrometers should be tuned regularly if their performance is to remain optimal and reliable. Indeed, the protocol of good laboratory practice (GLP) requires that mass spectrometers be tuned daily.

Tuning is accomplished by introducing a standard calibration compound into the instrument, then adjusting variables until the sensitivity and resolution are within acceptable limits. In modern GC/MS units this may be accomplished automatically by the data system with almost no input by the user. The most commonly used calibration standard is perfluorotri-*n*-butylamine [$(CF_3CF_2CF_2CF_2)_3N$; PFTBA], which gives fragment ions over nearly the entire mass range used for most routine GC/MS work (*m/z* 30 to 600; see Problem 1.1). Prominent ions at *m/z* 69, 219, and 502 are most often used for adjusting settings for instrument variables. Other calibration standards for higher masses are available, and some laboratories prefer to tune manually using a compound that is frequently encountered during their analyses.

Appropriate sensitivity is achieved by adjusting both the voltage (gain) applied to the electron multiplier, which directly affects the intensity of the signal output, and voltages to various components in the ion source. The latter adjustments are made because the metal surfaces of the source change somewhat each time a sample is run. As more and more ions are formed, then electrically neutralized on the source surfaces, polymeric organic deposits build up, causing local variations in the intense electric fields present in the source and thus interfering with ionization and the movement of ions out of the source and into the analyzer. Tuning helps counteract the effects of these deposits. After long-term use, the ion source surfaces become so dirty that even tuning does not provide the necessary remedy. At that point the source has to be removed from the instrument, disassembled, and cleaned. Ironically, clean sources may need to be tuned more often than dirty ones since the amount of effective deposit on the surfaces increases dramatically at first, then tapers off after several samples have been run.

Achieving good resolution and peak shape is more complex and cannot be done without some loss of sensitivity—obviously the more restrictive the conditions for allowing ions through the analyzer, the fewer the number of ions that will be allowed through. Variables in both the ion source and the analyzer affect the resolution and peak shape (Section 1.5.2), and all of these must be adjusted against each other in order to get both acceptable resolution and sensitivity.

Once well-resolved ions of reasonable intensity are obtained, the mass of each ion must be accurately determined. It is easy to forget that the mass spectrometer/computer combination really is not very "intelligent"—that is, the mass spectrometer can only provide information as either voltages or electric currents that the computer has

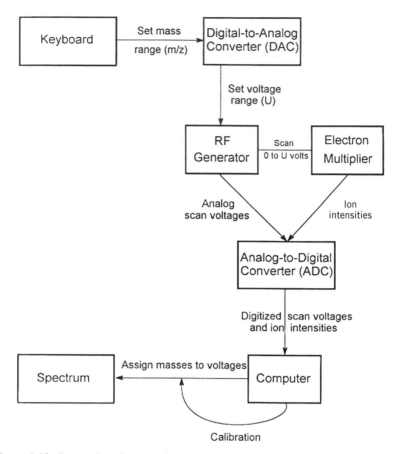

**Figure 1.12.** Interactions between the mass spectrometer and computer during a scan.

to interpret through its software. Conversely, the spectrometer is essentially lifeless without intelligent direction. The interactions between the mass spectrometer and computer during one mass spectral scan are shown in Figure 1.12.

The user, by means of the computer keyboard (or other appropriate input device), tells the spectrometer what mass range to scan and at what rate to collect the spectrum. This information is stored in binary (digital) form by the computer as a series of 0's and 1's (essentially on/off switches) corresponding to increasing powers of 2. For example, the number 27 is stored as 00011011: $0 \cdot 2^7 + 0 \cdot 2^6 + 0 \cdot 2^5 + 1 \cdot 2^4 + 1 \cdot 2^3 + 0 \cdot 2^2 + 1 \cdot 2^1 + 1 \cdot 2^0 = 0 + 0 + 0 + 16 + 8 + 0 + 2 + 1 = 27$. The voltage circuitry in the RF and dc generators, however, does not respond to digital information, so that the digital input first must be converted to analog form that the analyzer can use. In contrast to digital information, which occurs only in discrete integral steps (in the example above, the next available value is 00011100 or 28), analog information may have any measurable value (e.g., 27.0236). This conversion is accomplished by means of a digital-to-analog converter (DAC), an electronic circuit that takes the digital values provided by the computer and "steps" the voltage circuitry until the appropriate voltage value is reached. As an example, if we scan from $m/z$ 35 to 400, the dc generator might scan from 8 to 110 V. If the computer designates $m/z$ 35 as the 350th step and $m/z$ 400 as the 4000th step (assume for simplicity that 1 step = 0.1 dalton), the DAC redefines this scan as the 350th voltage step through the 4000th voltage step, with an individual voltage step equal to 110 - 8 V/(4000 - 350)steps = 102 V/ 3650 = 0.0279 V. Additional information can tell the voltage circuitry how long to stay at each voltage step, thus determining the rate of the scan. In our example a 0.5-s scan would take 0.5/3650 = 0.14 ms per voltage step.

During the course of a typical GC/MS run, the ion source is turned on by having the computer, at a time determined by the user, cause a current to flow through the filament. The analyzer then scans repeatedly through the assigned voltage range. Throughout this time the electron multiplier produces a variable signal current output, the magnitude of which depends on what ions are allowed to pass through the analyzer and how many of each of these ions there are. The analog (continuously variable) signal current produced by the multiplier is not directly usable by the computer, and thus must first be digitized by passage through an analog-to-digital convertor (ADC). If the voltages applied to the analyzer vary linearly with time (i.e., the voltage steps are all of equal duration), the signal current output of the multiplier can be correlated directly with the voltages being scanned by the analyzer.

If the masses being scanned were related to the voltages in a completely linear fashion (as in theory they should be for quadrupoles and ion traps; Figure 1.13a), the computer would have no difficulty assigning $m/z$ values to each of the voltages, and thereby could produce the mass spectrum from the voltages and multiplier signal currents alone. However, the ability of the system to produce this linearity between voltages and masses over time is less than ideal (Figure 1.13b; the deviations in this figure are purposely exaggerated), so that the analyzer voltages must be calibrated to correspond to known masses if the computer is to assign them correctly. With magnetic sector instruments, the relationship between $m/z$ and the scanned variable is not linear (Section 1.3.1), making calibration even more critical.

Calibration is accomplished by obtaining the spectrum of a known standard, usually the same PFTBA used for tuning. The known masses of the fragment ions of PFT-BA are preprogrammed into the computer, along with a range of probable analyzer voltages that will produce these masses. The computer then compares the observed peaks with those expected for PFTBA and assigns the corresponding voltages (or values for another variable) to the correct masses. Between the voltages for the observed fragment ions of PFTBA, the computer has to interpolate, but fortunately, the deviation from linearity is not terribly large. Although instruments should be calibrated on a daily basis to ensure correct mass assignments, quadrupole mass spectrometers are very stable and may hold nearly the same calibration for weeks or months at a time.

## 1.5.2. The Mass Spectrum

As we have just seen, the current output of the electron multiplier changes with time as ions of different $m/z$ values travel through the mass analyzer. Further, the $m/z$ val-

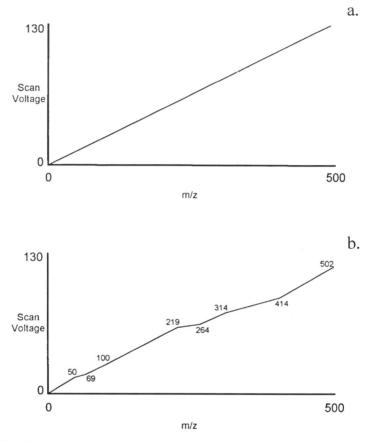

**Figure 1.13.** Mass spectrometers must be calibrated since the actual relationship between voltage and $m/z$ is nonlinear. (*a*) Theoretical scan of voltage vs. mass. (*b*) The actual scan is nonlinear; the computer matches known masses of PFTBA with observed voltages and interpolates between them.

ues found within the analyzer at any time are related to the value of the scanned variable (e.g., the voltages applied to the quadrupole rods) at that moment. But what does a mass spectral "peak" really "look like" to the electron multiplier? Consider, as an example, the molecular ion for a compound having a molecular weight of 200. At first we might suspect that, at the precise instant during the scan when the analyzer voltages allow ions of $m/z$ 200 to pass through to the detector, all of the $m/z$ 200 ions rush through and collide with the multiplier surface at exactly the same time.

When $\Delta M \sim 1$, however, as it is with most quadrupole analyzers, not all of the $m/z$ 200 ions traverse the analyzer at the same instant. Instead, because a small range of $m/z$ values is allowed through the analyzer at any given time under these conditions, a few $m/z$ 200 ions will begin to "leak" through the analyzer when the value of the applied voltage (or other variable) corresponds to about $m/z$ 199.5. The number of ions will increase as the value of this variable approaches that corresponding to *m/z* 200.0, then taper off again as it approaches that corresponding to *m/z* 200.5. If *m/z* 200 ions pass through the analyzer at lower or higher values, they will overlap with the passage of *m/z* 199 or 201 ions, and the mass resolution will be even lower. The resulting "peak" in this case is thus a curve similar to a typical GC peak, having a maximum value at approximately *m/z* 200.0 (Figure 1.14). The actual value of the maximum may differ from *m/z* 200.0 for reasons discussed in Section 2.1, but will rarely be lower than about 199.8 or greater than about 200.3.

In order for the data system to translate this information into a mass spectral peak, it must identify a multiplier signal current maximum during the time when analyzer variables permit the passage of ions having *m/z* values from 199.5 to 200.5. In practice this "window" of allowable *m/z* values is programmed to be from about 199.7 to 200.7 since, on the average, the actual current maximum will occur at values slightly greater than 200.0 (Section 2.1.2). The data system then assigns two values to these data—one identifying the *m/z* value, the other quantifying the intensity of the maximum signal current produced by the electron multiplier during that window. Some data systems assign only the nearest integral *m/z* value to the maximum, so that any

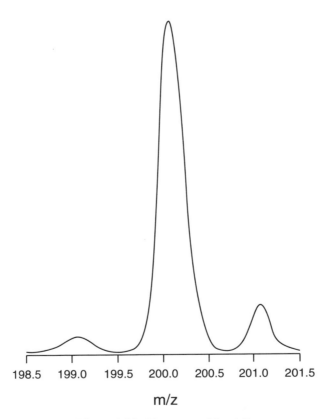

198.5    199.0    199.5    200.0    200.5    201.0    201.5

m/z

**Figure 1.14.** Mass spectral "peak."

**Table 1.3. Tabulated Mass Spectrum of Δ⁹-THC**

| m/z | Abundance | m/z | Abundance | m/z | Abundance | m/z | Abundance |
|---|---|---|---|---|---|---|---|
| 36.85 | 0.19 | 64.05 | 0.67 | 86.15 | 0.39 | 108.10 | 6.31 |
| 39.05 | 7.57 | 65.05 | 4.25 | 88.15 | 0.25 | 109.10 | 2.75 |
| 41.05 | 33.58 | 66.05 | 1.30 | 89.05 | 1.31 | 110.00 | 0.32 |
| 42.15 | 2.39 | 67.05 | 9.27 | 91.05 | 11.27 | 111.10 | 0.39 |
| 43.05 | 26.97 | 69.05 | 8.03 | 92.05 | 1.92 | 112.00 | 0.25 |
| 44.05 | 1.11 | 70.15 | 0.66 | 93.05 | 4.96 | 113.10 | 0.58 |
| 44.95 | 0.17 | 71.05 | 1.73 | 94.05 | 1.11 | 115.00 | 6.72 |
| 50.05 | 0.35 | 72.05 | 0.19 | 95.05 | 6.88 | 116.10 | 1.68 |
| 51.05 | 2.01 | 75.05 | 0.36 | 96.05 | 0.84 | 117.10 | 2.46 |
| 52.05 | 1.09 | 77.05 | 8.67 | 97.05 | 0.57 | 118.10 | 0.69 |
| 53.05 | 5.51 | 78.05 | 2.54 | 98.05 | 0.32 | 119.10 | 4.39 |
| 54.15 | 0.89 | 79.05 | 6.30 | 99.15 | 0.35 | 120.10 | 1.48 |
| 55.05 | 9.98 | 80.15 | 1.42 | 100.75 | 1.25 | 121.10 | 4.44 |
| 56.05 | 1.51 | 81.05 | 9.77 | 102.15 | 0.82 | 122.10 | 2.77 |
| 57.05 | 1.81 | 82.05 | 1.19 | 103.00 | 2.33 | 123.10 | 3.40 |
| 58.05 | 0.27 | 83.05 | 1.37 | 104.10 | 0.89 | 124.10 | 1.32 |
| 59.05 | 0.67 | 84.15 | 0.49 | 105.00 | 5.65 | 125.00 | 0.26 |
| 62.05 | 0.23 | 85.15 | 0.29 | 106.10 | 1.61 | 126.10 | 0.40 |
| 63.05 | 1.03 | 86.15 | 0.25 | 107.10 | 6.31 | 127.10 | 2.24 |
| 128.10 | 5.62 | 148.10 | 1.03 | 167.10 | 0.95 | 186.05 | 1.16 |
| 129.00 | 3.63 | 149.10 | 1.82 | 168.10 | 0.98 | 187.05 | 5.57 |
| 130.00 | 1.00 | 150.00 | 0.81 | 169.10 | 1.54 | 188.05 | 2.34 |
| 131.10 | 2.73 | 151.10 | 0.92 | 170.10 | 0.75 | 189.15 | 1.65 |
| 132.10 | 0.91 | 152.10 | 2.92 | 171.05 | 1.91 | 190.15 | 1.21 |
| 133.10 | 2.09 | 153.10 | 2.44 | 172.05 | 1.02 | 191.15 | 1.35 |
| 134.00 | 1.43 | 154.10 | 1.12 | 173.05 | 3.62 | 193.05 | 8.60 |
| 135.10 | 1.97 | 155.10 | 1.47 | 174.05 | 8.21 | 194.15 | 1.40 |
| 136.10 | 1.22 | 156.10 | 0.83 | 175.05 | 4.60 | 195.05 | 1.33 |
| 137.00 | 1.40 | 157.10 | 2.23 | 176.05 | 1.89 | 196.15 | 0.73 |
| 138.10 | 0.43 | 158.10 | 1.36 | 177.05 | 1.07 | 197.05 | 1.90 |
| 139.00 | 0.57 | 159.10 | 2.27 | 178.05 | 1.29 | 198.05 | 1.09 |
| 141.10 | 3.38 | 160.10 | 2.36 | 179.05 | 0.76 | 199.05 | 3.23 |
| 142.10 | 1.47 | 161.10 | 3.51 | 180.05 | 0.53 | 200.05 | 2.08 |
| 143.10 | 1.51 | 162.10 | 0.90 | 181.05 | 1.88 | 201.05 | 6.42 |
| 144.00 | 1.41 | 163.10 | 1.56 | 182.05 | 1.09 | 202.15 | 2.19 |
| 145.10 | 2.76 | 164.10 | 0.63 | 183.05 | 1.78 | 203.15 | 1.07 |
| 146.10 | 0.82 | 165.10 | 3.60 | 184.05 | 0.79 | 204.15 | 0.30 |
| 147.10 | 4.39 | 166.10 | 1.29 | 185.05 | 1.95 | 205.05 | 0.55 |
| 206.25 | 0.32 | 224.15 | 0.45 | 246.10 | 3.41 | 274.20 | 0.21 |
| 207.15 | 1.70 | 225.15 | 1.03 | 247.10 | 0.67 | 281.20 | 0.52 |
| 208.05 | 0.43 | 226.15 | 0.59 | 253.10 | 0.31 | 282.20 | 0.20 |
| 209.05 | 0.55 | 227.15 | 1.57 | 254.20 | 0.23 | 283.20 | 0.72 |
| 210.15 | 0.54 | 228.15 | 1.34 | 255.10 | 1.09 | 284.20 | 0.43 |
| 211.05 | 1.42 | 229.15 | 2.59 | 256.20 | 1.08 | 285.20 | 3.03 |

**Table 1.3.** (*Continued*)

| m/z | Abundance | m/z | Abundance | m/z | Abundance | m/z | Abundance |
|-----|-----------|-----|-----------|-----|-----------|-----|-----------|
| 212.15 | 0.95 | 231.15 | 46.22 | 257.20 | 7.02 | 286.20 | 2.16 |
| 213.05 | 2.78 | 232.15 | 10.54 | 258.20 | 21.89 | 287.20 | 0.49 |
| 214.15 | 1.93 | 233.15 | 4.27 | 259.10 | 4.62 | 295.20 | 0.30 |
| 215.15 | 3.88 | 234.15 | 0.55 | 260.10 | 0.71 | 297.20 | 5.96 |
| 216.15 | 1.15 | 239.15 | 0.36 | 267.10 | 0.25 | 299.20 | 100.00 |
| 217.15 | 6.79 | 240.10 | 0.50 | 268.20 | 0.19 | 300.20 | 22.01 |
| 218.15 | 1.47 | 241.10 | 1.16 | 269.10 | 0.96 | 301.20 | 2.46 |
| 219.15 | 0.88 | 243.10 | 27.41 | 271.20 | 38.45 | 314.15 | 70.52 |
| 221.15 | 3.39 | 244.10 | 6.00 | 272.20 | 9.38 | 315.25 | 16.83 |
| 222.15 | 0.69 | 245.10 | 3.21 | 273.20 | 1.52 | 316.15 | 2.46 |
| 223.25 | 0.30 | — | — | — | — | — | — |

peak in the *m/z* 199.7 to 200.7 window would be labeled *m/z* 200. Others report the maximum more accurately, sometimes to the nearest 0.05 dalton.

The mass spectrum, then, consists of the collection of *m/z* values and corresponding values for the maximal multiplier currents from the windows defined by those *m/z* values, over the entire range of *m/z* values scanned by the analyzer. It is customary to assign the value of 100% to the largest of all of the current maxima obtained during an individual scan and to report the remaining values relative to that figure. The largest ion in any mass spectrum (100% relative intensity) is called the *base peak*. Although it is easy to think of the base peak as fixed for the mass spectrum of any given compound, it is in fact dependent upon the depicted mass range. For example, if the spectrum of an intermediate molecular weight, primary aliphatic amine is scanned (and reported) from *m/z* 10 to 300, the base peak in the spectrum will nearly always occur at *m/z* 30 (Section 5.3). If the spectrum is only scanned from *m/z* 35 to 300, however, as many analytical labs with quadrupole GC/MS units do, the ion at *m/z* 30 will not be reported, and some other ion in the spectrum will become the base peak.

Graphically, *m/z* values are plotted along the *x* axis and relative abundances or intensities along the *y* axis. The mass spectra shown in the figures in this book will be in this form. The same data may be presented in tabular form, such as that in Table 1.3. You may notice that not every *m/z* value is represented in Table 1.3 or in any other mass spectrum in this book. In fact, no organic compound can produce fragment ions at every mass. In addition, to minimize the number of spurious "noise" peaks that occur in all spectra, the data system applies a threshold value for multiplier output below which peaks will not be reported. In Table 1.3 the threshold appears to be approximately 0.1% of the size of the most intense ion.

### 1.5.3. Library Searches

Although the computer performs a number of data reduction and manipulation procedures, the identification of an unknown mass spectrum by comparison with those in collections stored in the computer is one of the most useful. This "library search" is a very powerful tool since it accomplishes in a few seconds what might take the operator more than several hours to perform manually. Yet, precisely because of its perceived power, the library search can be overemphasized. A brief explanation of how library searches work should help put their results in better perspective.

Most computer libraries contain only condensed spectra—that is, they retain only the most meaningful (unique) peaks in each spectrum. The number of peaks retained per spectrum usually varies from 10 to 50, depending on the search program. This is done so that the search algorithm (the mathematical process by which the search program compares spectra) does not have to compare the unknown spectrum peak for peak with the entire spectrum of each library entry, a task that would use a lot of computer time. In practice, very little is sacrificed by using condensed spectra.

Both library and unknown spectra are condensed using a weighting factor. A common weighting factor is the square root of the mass, so that

$$\text{Weighted intensity} = \text{observed intensity} \times (\text{mass})^{\frac{1}{2}}$$

This factor is appropriate since, as we shall see in subsequent chapters, ions occurring at high mass are more likely to be characteristic of the compound in question than those at low mass. In another approach, an algorithm called probability based matching (PBM), developed by Prof. Fred McLafferty of Cornell University, assigns individual uniqueness and abundance factors to various peaks in each spectrum. Like the weighting factor above, the uniqueness of the ion increases logarithmically with its mass. Each spectrum is stored as a list of $m/z$ values (usually only as integral values) along with the weighted intensity associated with each mass.

The library may be subjected to a presearch, so that only a small percentage of the total number of spectra in the library need be compared thoroughly. This also saves computer time without sacrificing accuracy. In the presearch, only a few of the most important peaks in the unknown are compared with a few of the most important peaks in each library spectrum. For example, unknown spectra that have several intense high mass ions (and thus a high degree of uniqueness) may be vaguely similar to only a dozen or so spectra in the library. Library spectra that show no promise of providing matches are excluded from the rest of the search.

The remainder of the search is performed in one of two ways—as a forward search or as a reverse search. The forward-search algorithm compares the weighted condensed unknown spectrum with the similarly weighted condensed subset of library spectra selected during the presearch. This comparison is made by assigning to each ion a vector starting at the origin and having, for example, the mass as the $x$ coordinate and the weighted intensity as the $y$ coordinate. The vectors for all of the ions in the unknown are added and the result compared with a similarly summed vector generated from the peaks in each library spectrum. The relative degree of "match" between these vectors determines the relative similarity between the spectra.

In a reverse search, each spectrum in the condensed library subset is compared with the unknown spectrum. In PBM searches, the full, rather than condensed, unknown spectrum is used, to account for the possibility that the unknown mass spectrum might actually be a mixture of spectra. The relative abundance of each ion in the library spectrum is compared with the corresponding ion in the unknown. If several ions in the spectra do not compare favorably in intensity, the algorithm moves on to the next spectrum. When spectra having similar characteristics are found, a factor indicating the probability of match is calculated based on the uniqueness and intensities of the ions that have matched. If the data system has access to more than one spectral library, the algorithm may be set up to search all of the libraries sequentially and provide a composite list of search results.

Library search programs have several important strengths: (a) they give immediate access to up to 150,000 or more spectra and compare the unknown to these spectra in a small fraction of the time it would take to compare the spectra visually; (b) most algorithms tend to rely only on local relative abundances within each spectrum, so that spectra obtained under somewhat different conditions, or on different instruments, usually do not affect the search results[2]; (c) they may reveal the identity of the unknown specifically by a direct match, or, in cases where the unknown compound is not in the library, may at least give some hints as to the general structure; and (d) they may identify compounds even when they are present in mixtures, as often happens when compounds coelute during the chromatography of complex samples such as plant extracts or gasoline residues.

On the other hand, the importance of library search results can be seriously overemphasized. Finding a high correlation (high "match index" or "probability of match") between an unknown spectrum and a library spectrum does not necessarily mean that the unknown has been identified unequivocally. This point cannot be stated too emphatically. **The criteria necessary to identify an unknown by mass spectrometry must include at least a visual comparison of the unknown and library spectra by the analyst** and may also demand additional information such as a comparison of GC retention times.

There are several reasons why this is so. First, similar structures often give spectra that are not easily distinguished by library search (or even visually, for that matter). Optical isomers cannot be distinguished at all, and other stereoisomers may be distinguishable only with difficulty after repeated scans determine which minor differences are real and which are not. Even some positional isomers (especially aromatic hydrocarbons; see Figure 1.15) may not be identified with complete certainty by mass spectrometry alone. Second, some types of search algorithms are more discriminating against certain types of compounds than are others. The PBM algorithm, in fact, has such difficulty discriminating between certain aliphatic amines (ephedrine, methamphetamine, and amphetamine among them) that the "correct match" may not appear at the top of match list, if it appears at all (Figure 1.16). Even with spectra containing several "characteristic" high mass ions, the PBM search may fail to produce even a viable correct match candidate under some circumstances (Figure 1.17).

Conversely, the correct match for the unknown may not be in the library, even though the search algorithm may show strong correlations between the unknown and other spectra in the library. Further, if the spectrum is not "clean" (i.e., numerous ions from extraneous materials are present), the search results may reflect the presence of

---

[2]Although most search algorithms will correctly identify spectra from different instruments or obtained under different conditions, the best matches will be attained from spectra run on *your instrument under your conditions.*

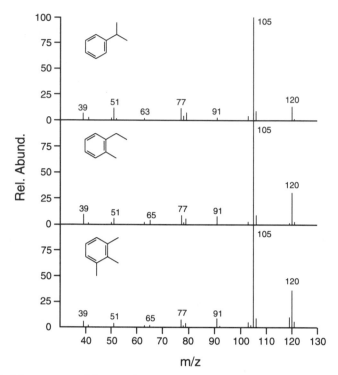

**Figure 1.15.** Mass spectra of these positional isomers are so similar that they may not be distinguished by library search.

the extraneous ions rather than those of the unknown. Finally, although the publishers of mass spectral databases are usually very careful about which spectra to include in their libraries, errors do occur with enough frequency that it should give any user pause. *Never, ever, just assume that the library spectrum is correct!* Indeed, one of the purposes of this book is to give you tools that will help you evaluate whether a given spectrum reflects the assigned chemical structure or not.

### 1.5.4. Using the Data System to Analyze GC/MS Data

Even prior to 1980 (when the first truly sophisticated data systems began to appear commercially), the combination of mass spectrometry and gas chromatography revolutionized the analysis of complex mixtures. Yet most younger mass spectroscopists can hardly imagine a GC/MS unit *not* attached to a personal computer loaded with data collection and manipulation software. Since many of the readers of this book will (or already do) work in laboratories where GC/MS is a primary method of sample analysis, it is worthwhile to explore how the data system can be used in the analysis of sample mixtures. Alternative data handling for some highly complex mixtures is given in Section 4.2.3.

As an example, consider an unknown solid brought to a forensic laboratory for illicit drug analysis. To minimize sample workup, this solid was dissolved directly in

PBM Search of library C: \DATABASE\DRUGS.L

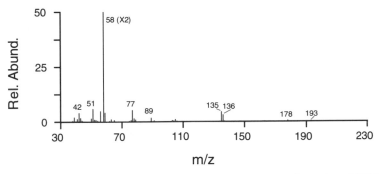

| | Name | MolWT | Formula | Qual |
|---|---|---|---|---|
| 1. | Phenol, 4–[2–(methylamino) propyl– | 165 | C10H15NO | 40 |
| 2. | Ephedrone (Methcathinone) | 163 | C10H13NO | 9 |
| 3. | N–Methyl–3, 4–Methylenedioxyamphetamine | 193 | C11H15NO2 | 9 |
| 4. | 1–Phenyl–2–butanamine | 149 | C10H15N | 9 |
| 5. | Butanedioic acid, salt with N, N–dimeth | 388 | C21H28N2O5 | 4 |

**Figure 1.16.** Mass spectrum of *N*-methyl-3,4-methylenedioxyamphetamine (MDMA) obtained during analysis of an illicit drug sample. The correct match is not listed first in the search results, in spite of the fact that the spectrum matches well with the standard (Figure 8.2*b*). All of the compounds listed in the search results have intense *m/z* 58 ions.

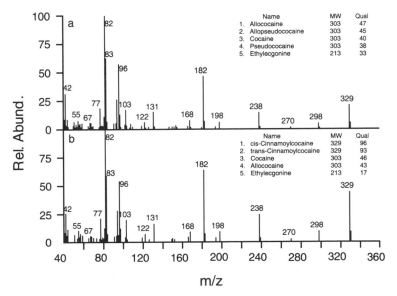

| | Name | MW | Qual |
|---|---|---|---|
| 1. | Allococaine | 303 | 47 |
| 2. | Allopseudococaine | 303 | 45 |
| 3. | Cocaine | 303 | 40 |
| 4. | Pseudococaine | 303 | 38 |
| 5. | Ethylecgonine | 213 | 33 |

| | Name | MW | Qual |
|---|---|---|---|
| 1. | cis-Cinnamoylcocaine | 329 | 96 |
| 2. | trans-Cinnamoylcocaine | 329 | 93 |
| 3. | Cocaine | 303 | 46 |
| 4. | Allococaine | 303 | 43 |
| 5. | Ethylecgonine | 213 | 17 |

**Figure 1.17.** Mass spectra of *cis*-cinnamoylcocaine obtained from two separate, but otherwise similar, samples. The correct match does not even appear in the search results for the upper spectrum, while the lower spectrum is correctly identified by the same search algorithm.

methanol, and 1 μL of this solution was injected onto a 10 *m* phenylmethylsilicone (HP-1) capillary column. The GC oven was programmed (via the data system) to start the run at 160°C, heat from 160 to 280°C over a 6-min period, then remain at 280°C for an additional minute. During the same time interval, the mass spectrometer was programmed (a) to keep the filament and electron multiplier *off* for 1 min while the methanol was passing through the GC column (see Sections 1.2 and 1.4); (b) to scan repeatedly from *m/z* 35 to 350 at a scan rate of about 0.5 s/scan for the next 3 min; and (c) to scan repeatedly from *m/z* 35 to 400 for the remainder of the run. Keeping the scan range short during the early part of the GC run shortened individual scan times, thus reducing the loss of chromatographic resolution. Later in the run, chromatographic resolution is often less important (especially for isothermal runs), but it becomes important instead to scan to a high enough mass that no molecular ions are inadvertently missed.

By scanning continuously in this way, the MS also acts as a GC detector. In fact, one way of reporting GC/MS data is as a *total ion chromatogram,* or *TIC,* which is a plot of the total output of the electron multiplier per mass spectral scan over the time of the run. In the TIC for this sample (Figure 1.18), the electron multiplier output is expressed in terms of total ion abundances, that is, the total number of ions produced in each mass spectral scan (which is directly proportional to the total current produced

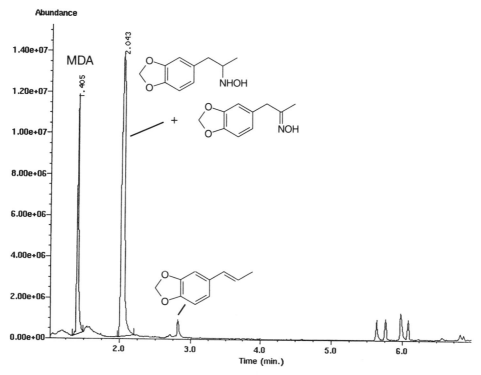

**Figure 1.18.** Total ion chromatogram (TIC) generated from mass spectral data collected during analysis of an illicit drug sample.

by the multiplier during the scan; Section 1.4). Although this plot looks like that produced by a flame ionization detector (FID), there is one subtle (but sometimes important) difference: the sensitivity of the MS as a GC detector may vary widely from compound to compound, depending on how efficiently each molecule fragments and how many fragment ions are produced. This variability exceeds that normally encountered with FID detectors (although not with some other types of GC detectors) and may lead to situations where the same sample mixture shows noticeably different chromatograms by GC (FID) and MS.

Since the data system has stored the data from the run as a collection of mass spectral scans, each of these spectra can be retrieved, thereby identifying the makeup of the GC effluent at any given time. With most modern software packages, spectrum retrieval is accomplished simply by moving the cursor on the monitor to the point on the chromatogram we wish to investigate, then pressing "Enter" on the keyboard or by clicking a mouse button. For single-component GC peaks, an acceptable spectrum can be obtained either by placing the cursor at the top of the chromatographic peak or by asking the data system to average the spectra from several scans over the top of the peak (Section 1.3.5). Background that is reasonably constant over the peak can be subtracted from this spectrum by the data system, by choosing a point near the front (or back) base of the peak, or by asking the data system to average background spectra from the same area(s). Background subtraction may not be necessary with intense GC peaks and minimal column bleed.

Such a process produced the spectra in Figure 1.19*a* and *d,* which were identified by library search (*and visual confirmation by the analyst*) as 3,4-methylenedioxyamphetamine [retention time (r.t.) 1.405; MDA, a controlled hallucinogenic drug—Section 8.2] and 1-(3,4-methylenedioxyphenyl)propene (r.t. 2.8), an intermediate produced during one of the common syntheses of MDA and its analogs. A similar treatment of the large (and broad) peak at r.t. 2.04 led to a spectrum that was much like that shown in Figure 1.19*c,* except for the addition of a few small ions (most notably one at $m/z$ 195). This compound was identified by library search as 1-(3,4-methylenedioxyphenyl)-2-propanone oxime, which, based on the analyst's experience, was an unexpected result. Further examination of individual spectra over this peak indicated the presence of two different compounds that were not entirely separated chromatographically.

Obtaining an acceptable spectrum for each of these compounds (one that was devoid of peaks from the other) necessitated finding a method of determining the elution time of each of the contributors (see Section 1.3.5). Although mathematical deconvolution can be done even in the absence of MS data, the data already collected in this case contained enough information to obtain the desired result. As it turns out (Figure 1.19*b* and *c*), the mass spectra of the two coeluting compounds each contained a few intense ions not present in the other. In particular, the base peak ($m/z$ 60) in the spectrum of the first-eluting compound, and the relatively abundant molecular ion ($m/z$ 193) in the spectrum of the second, were characteristic enough of each compound that they could be used to differentiate the two. Thus, instead of asking the data system to plot *total* ion abundances over the course of the run (the TIC), we instructed it to plot only the abundances of the two characteristic ions. The results of this process, known by a variety of names including *reconstructed ion chromatography*

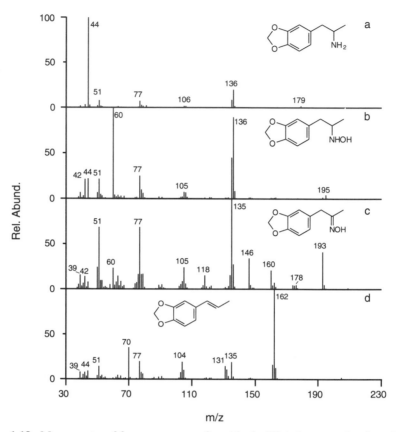

**Figure 1.19.** Mass spectra of four components found in the illicit drug sample whose TIC is shown in Figure 1.18. (a) 3,4-Methylenedioxyamphetamine (MDA; r.t. 1.40 min); (b) N-hydroxy-MDA (r.t. 2.02 min); (c) 1-(3,4-methylenedioxyphenyl)-2-propanone oxime (r.t. 2.05 min); and (d) 1-(3,4-methylenedioxyphenyl)propene (r.t. 2.81 min).

*(RIC)* and mass chromatography, are shown in Figure 1.20 for the small window of time between r.t. 1.75 and 2.27 min. Reconstructed ion chromatography differs from selected ion monitoring (SIM; Section 1.3.2) in that the RIC is generated *after collecting complete mass spectral scans* for the entire chromatogram; in SIM, ions must be selected prior to sample injection. Thus, an RIC can be generated for any ion in the collected mass range, whereas no data is available in SIM for any ions other than those selected before the run begins.

Figure 1.20 makes clear that, although the two compounds elute only 0.03 min (less than 2 s!) apart and with a considerable amount of overlap, it should be possible to obtain spectra of both compounds that have little contamination due to ions from the other. Thus the spectrum of N-hydroxy-3,4-methylenedioxyamphetamine (N-hydroxy-MDA; Figure 1.19b) was obtained from the upper front side of the peak in the *m/z* 60 chromatogram, at a point corresponding to the base of the peak in the *m/z* 193 chromatogram, while the spectrum of the oxime (Figure 1.19c) was obtained

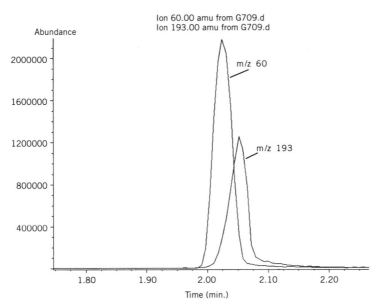

**Figure 1.20.** Reconstructed ion chromatograms (RICs) for $m/z$ 60 and 193 generated from the data collected during the run shown in Figure 1.18. These chromatograms help locate acceptable spectra for each of the two coeluting components.

at a point just beyond the top of the $m/z$ 193 peak where the first compound had stopped eluting.

It is tempting to conclude, based on the relative sizes of the peaks in the two RICs, that N-hydroxy-MDA is the more concentrated of the two coeluting components. Actually, without doing further analytical work, this cannot be determined on the basis of these data. First of all, the $m/z$ 60 ion in the N-hydroxy-MDA spectrum is the base peak, while $m/z$ 193 in the spectrum of the oxime has a relative abundance of only about 45%. Thus, we *expect* the two RICs to have different intensities. In addition, we know very little about how difficult it is to ionize and fragment each of these compounds (see above), so that a comparison of concentrations based on the RIC data alone becomes artificial.[3]

## 1.6. CRITERIA FOR GOOD QUALITY SPECTRA

Mass spectra cannot be interpreted if they contain *mis*information. Unfortunately, even refereed journals and carefully edited collections of standard spectra contain spectra that, from time to time, fail to meet this criterion. Judging whether or not a mass spectrum is credible is sometimes the most critical step in its interpretation.

[3]For those who are curious, N-hydroxy-MDA apparently disproportionates to MDA and the oxime in the injection port of the GC, thus accounting for the presence of these three compounds in the sample. Analysis of the solid by infrared spectrometry indicates the presence of only N-hydroxy-MDA!

Several criteria are useful for evaluating the quality of spectra—not only those generated in our own laboratories but also those in the literature or under consideration for placement in user-generated libraries or for publication. These spectra

1. *should* show appropriate isotopic abundance ions (e.g., should not be missing $^{13}C$ isotope peaks), especially at high mass (see Chapter 2);
2. *should* not exhibit excessive background noise (spurious peaks of instrument origin throughout the mass range) or the obvious presence of extraneous materials (e.g., column bleed, coeluting GC peaks, contaminants in the ion source); and
3. *must* be consistent with the known or proposed structure—that is,
   a. all masses must be correct;
   b. the molecular ion, if present, must correspond to the mass of the molecular weight;
   c. neutral losses at high mass must reflect functional groups or structural arrangements present in the molecule (Chapter 4); and
   d. the base peak must be consistent with the structure (Chapters 4 to 6).

Although these criteria seem to reflect nothing more than common sense, they are violated just often enough to warrant special mention. You may want to review this section after solving some mass spectral unknowns and studying the material contained in later chapters.

## 1.7. PROBLEMS

Although answers to all of the problems are given at the end of this book (Chapter 9), you are strongly encouraged to attempt to solve each problem before proceeding further. In my experience, there is no substitute for solving unknowns if you want to be-

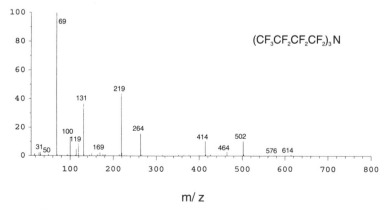

**Figure 1.21.** Mass spectrum of perfluorotributylamine (PFTBA).

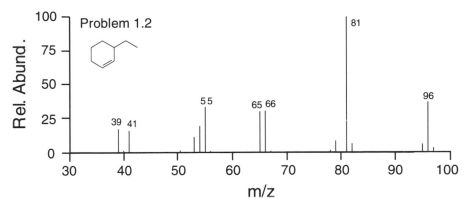

**Figure 1.22.** Library spectrum for 3-ethylcyclohexene.

come proficient at interpreting mass spectra. I have consistently found that students who do not spend time working problems soon get lost, since most of the topics in this book build upon what has preceded them. Very few readers, unless they are already adept at mass spectral interpretation, will be able to absorb the concepts illustrated in the problems if they depend entirely on the explanations provided.

***Problem 1.1.***  The compound commonly used for tuning and calibration of mass spectrometers is PFTBA, a trade name for perfluorotri-*n*-butylamine, $(CF_3CF_2CF_2CF_2)_3N$. This compound exhibits peaks of at least moderate intensity over the entire mass range normally used in GC/MS work (Figure 1.21). The most prominent peaks in the spectrum of PFTBA occur at *m/z* 31, 69, 100, 114, 119, 131, 219, 264, 414, 464, 502, 576, and 614. Devise empirical formulas (and, if possible, hypothesize structures) for each of these ions.

***Problem 1.2.***  The spectrum in Figure 1.22 for 3-ethylcyclohexene, showing a molecular ion at *m/z* 96 and a base peak at *m/z* 81, is from a published mass spectral library collection. What is wrong with this spectrum?

# CHAPTER 2

# ISOTOPIC ABUNDANCES

## 2.1. NATURAL ISOTOPIC ABUNDANCES

The electron ionization mass spectrum of methane is shown in Figure 2.1. Interpretation of this spectrum seems straightforward, even for the novice. The molecular ion ($CH_4^{+}$), which is the base peak in the spectrum, corresponds to the molecular weight of methane (16 daltons).[1] The molecular ion consecutively loses a series of hydrogen radicals (H·) to give ions at $m/z$ 15 ($CH_3^{+}$), 14 ($CH_2^{+}$), 13 ($CH^+$), and 12 ($C^{+}$), whose relative intensities reflect what might be intuited about the relative stabilities of these ions.

It is easy to overlook the small peak at $m/z$ 17 in this spectrum. In fact, this peak is small enough that it might easily be ignored as just an artifact or an impurity. Repeated scanning under ideal conditions proves these explanations false, however. Provided the spectrometer is working properly, the peak at $m/z$ 17 is always there. Furthermore, the size of $m/z$ 17 relative to $m/z$ 16 is always approximately 1.1%, and the presence and size of this peak is generally independent of the type of mass spectrometer used.

More careful reasoning might suggest that this peak is due to a small amount of $CH_5^{+}$ formed by an ion–molecule reaction between the molecular ion and a neutral molecule of methane. Indeed, if this spectrum were measured using a high pressure (about 1 torr) of methane in the ion source (as occurs when methane is used as a reagent gas in CIMS—Section 1.2), this explanation would probably be correct. Under such conditions, in fact, $m/z$ 17 would probably become the base peak in the spectrum. However, at source pressures of $10^{-6}$ to $10^{-4}$ torr normally used for electron

---

[1]The $^{+}$, $+$, and · conventions for denoting the electronic state of ions and radicals will be discussed in Section 3.1.

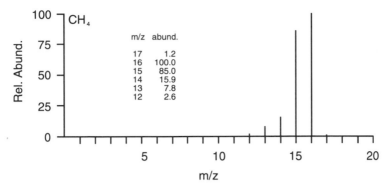

**Figure 2.1.** Electron ionization mass spectrum of methane.

ionization mass spectrometry, the intensity of the ion at $m/z$ 17 is independent of concentration over a fairly wide range, indicating that this ion is not formed via ion–molecule reactions.

So where does this peak come from? A clue about its origin comes from one of the ions observed in **Problem 2.1** (Figure 2.2), the spectrum of another carbon-containing compound. Try to figure out what this compound is before reading any further.

The Unknown in Problem 2.1, like methane, contains one carbon atom and, in addition, exhibits an ion one dalton higher in mass than the molecular ion with an intensity slightly greater than 1% relative to that of the molecular ion. Since these two molecules have nothing else in common, we might suspect that these peaks above the molecular ions originate from the presence of carbon in the molecule. This hypothesis turns out to be correct. But why is the size of these peaks about 1%?

Most elements occur naturally as mixtures of various isotopes—atoms of the same element that differ in atomic weight because, although they contain the same number of protons and electrons, they differ in the number of neutrons in the nucleus. A list

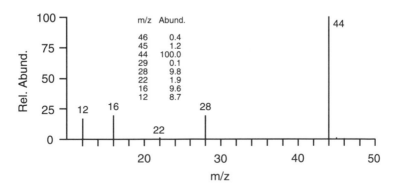

**Figure 2.2.** Mass spectrum for Problem 2.1.

**Table 2.1. Natural Isotopic Abundances of Some Common Elements**

| Isotope Mass | | Percent Abundance | Isotope Mass | | Percent Abundance |
|---|---|---|---|---|---|
| $^1$H | 1.008 | 99.99 | $^{28}$Si | 27.977 | 92.17 |
| | | | $^{29}$Si | 28.977 | 4.71 |
| $^{12}$C | 12.000 | 98.89 | $^{30}$Si | 29.974 | 3.12 |
| $^{13}$C | 13.003 | 1.11 | | | |
| | | | $^{32}$S | 31.972 | 95.03 |
| $^{14}$N | 14.003 | 99.64 | $^{33}$S | 32.971 | 0.76 |
| $^{15}$N | 15.000 | 0.36 | $^{34}$S | 33.968 | 4.20 |
| $^{16}$O | 15.995 | 99.76 | $^{35}$Cl | 34.969 | 75.77 |
| $^{18}$O | 17.999 | 0.20 | $^{37}$Cl | 36.966 | 24.23 |
| $^{19}$F | 18.998 | 100.00 | $^{79}$Br | 78.918 | 50.52 |
| | | | $^{81}$Br | 80.916 | 49.48 |

of some common elements and their naturally occurring isotopes is given in Table 2.1.

Table 2.1 introduces some important concepts. First, the abundances of the isotopes shown occur naturally; none are products of laboratory processes that manufacture man-made isotopes. For example, the carbon contained in diamonds recovered recently from a diamond mine would contain (in our solar system, at any rate) 98.9% carbon atoms that were $^{12}$C and 1.1% that were $^{13}$C. Outside our solar system these values appear to vary widely. It may seem surprising that $^{14}$C is missing from this list. Although it is a naturally occurring isotope of carbon (and forms the basis for $^{14}$C radioactive dating in archaeology), the natural abundance of $^{14}$C is so low that it is not normally observed in electron ionization mass spectrometry.

If we apply this idea to the spectrum of methane, we see that the molecular weight of $^{12}$CH$_4$ is 16 daltons (12 for the carbon and 1 for each of the hydrogens), while that of $^{13}$CH$_4$ is 17 (13 for the carbon and 1 for each of the hydrogens). Since ions are separated in mass spectrometry according to their mass-to-charge ratio, these two ions are readily distinguished in the mass spectrum. Indeed, mass spectrometry offers one of the best ways to identify and quantify the presence of different isotopes in samples. Further, the ratio of the two ions at $m/z$ 17 and 16 is directly related to the natural abundances of the two carbon isotopes [1.1% (for $^{13}$C) / 98.9% (for $^{12}$C) = 1.1%]. The same logic applies to the intensity of the $m/z$ 45 ion in the spectrum of the unknown in Problem 2.1.

### 2.1.1. Actual and "Average" Atomic and Molecular Weights

Table 2.1 underscores the distinction between the atomic weight of a designated isotope of an element (or the molecular weight of a compound having specified isotopes of all of the constituent elements) and the *average* atomic weight of an element as it exists in nature (or the average molecular weight of a compound derived from the average atomic weights of all of the constituent elements). For example, although the atomic weight of $^{12}C$ is 12.0000 daltons and that of $^{13}C$ is approximately 13.00, the "atomic weight of carbon" that appears in most literature sources is 12.01. This number is understandable when the natural abundances of $^{12}C$ and $^{13}C$ are factored in, that is,

$$\text{Ave. at. wt.} = [(\text{natural abundance of } ^{12}C)(12) + (\text{natural abundance}$$
$$\text{of } ^{13}C)(13)]/100\%$$
$$= [(98.9\%)(12) + (1.1\%)(13)]/100\%$$
$$= 12.01$$

For carbon, this difference is not large enough to cause serious errors for most compounds, since the average molecular weight of a compound having even 25 carbon atoms will only be 0.25 daltons higher than that of the same compound containing all $^{12}C$. The same is not true for some of the other elements, however. Consider the case of chlorine, whose two naturally occurring isotopes differ by not one, but two, daltons. In addition, $^{35}Cl$ accounts for only 75% of all natural chlorine; $^{37}Cl$ accounts for the rest. Using an equation like the one above (and using the precise natural abundances and atomic weights for $^{35}Cl$ and $^{37}Cl$ from Table 2.1), we can calculate the average atomic weight of chlorine:

$$\text{Ave. at. wt.} = [(\text{nat. abund. of } ^{35}Cl)(34.969) + (\text{nat. abund.}$$
$$\text{of } ^{37}Cl)(36.966)]/100\%$$
$$= [(75.77\%)(34.969) + (24.23\%)(36.966)]/100\%$$
$$= 35.453$$

This is the value given in the periodic table. Calculating the average molecular weight of the chlorine molecule ($Cl_2$) using this figure, we get a value of 70.91, or 71 to the nearest integral mass. This number is often found in references that list average molecular weights (and most references do!). This value, however, has no meaning in mass spectrometry since, because mass spectrometry allows us to observe individual isotopes, *we do not observe any peak at m/z 71* in the spectrum of $Cl_2$ ($^{35}Cl_2$ has a molecular weight of 70, $^{35}Cl^{37}Cl$ has a molecular weight of 72, and $^{37}Cl_2$ has a molecular weight of 74).

*Problem 2.2.* Using the data in Table 2.1, calculate the average atomic weight of bromine and the average molecular weight of $Br_2$. What peaks do you expect to see in the molecular ion region of the mass spectrum of $Br_2$?

### 2.1.2. Mass Defects

By standard definition, the atomic weight of $^{12}C$ is 12.0000 daltons. The atomic weights of all other isotopes of all the other elements are determined as ratios against this standard, which leads to nonintegral values. The difference between the actual atomic mass (relative to $^{12}C$) and the nearest integral mass is called the *mass defect*. The variation of mass defects over the periodic table is shown in Figure 2.3. Although mass defects for the very light elements are small and slightly positive, they are sub-

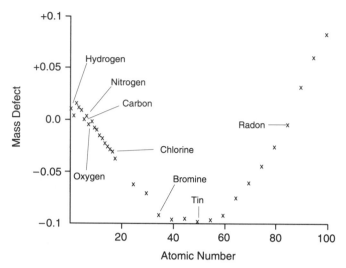

**Figure 2.3.** Mass defects for the elements.

stantially negative for the vast majority of elements. Only with the very heavy elements do values tend toward zero again and finally exceed it near radon (element 86).

Mass defects show up in calculations of the exact molecular weights of compounds. Consider $\Delta^9$-tetrahydrocannabinol (THC), the biologically active component of marijuana, which has an empirical formula of $C_{21}H_{30}O_2$. Using the data in Table 2.1 (which is more precise than that given in Figure 2.3), the mass defect (denoted by the Greek letter $\Delta$; note the same symbol is used in the naming of THC, where it denotes the position of an isolated double bond) for the molecular weight of THC is

$$\Delta = (12.000 - 12.000)(\times 21 \text{ carbons}) + (1.008 - 1.000)(\times 30 \text{ hydrogens}) +$$
$$(15.995 - 16.000)(\times 2 \text{ oxygens})$$
$$= (0.000)(21) + (0.008)(30) + (-0.005)(2)$$
$$= 0.000 + 0.240 - 0.010 = 0.230$$

Thus the actual molecular weight for $^{12}C_{21}{}^{1}H_{30}{}^{16}O_2$ is not exactly 314.000, but rather 314.230 because of the mass defects in the atomic masses of hydrogen and oxygen.

Mass defects are of little consequence if $m/z$ values are reported only to the nearest integer. However, with instruments that measure the mass-to-charge ratio more precisely, more information can be gleaned from this data. The instruments in our laboratory, for instance, report the molecular weight of THC as $314.2 \pm 0.1$ (see Table 1.3). Indeed, a closer examination of the $m/z$ values for all of the peaks in Table 1.3 shows that the masses of ions below about $m/z$ 140 have little or no mass defect, those from about $m/z$ 140 to 250 show an average mass defect of about 0.1 dalton, and those above $m/z$ 250 exhibit a mass defect of about 0.2 daltons. These numbers reflect in a general way the number of hydrogens in the ions since the mass defect of hydrogen makes the largest contribution to the overall mass defect in the molecular weight of THC.

In selected ion monitoring (Section 1.3.2), optimizing the analysis of a compound demands knowing an accurate molecular weight for each ion selected. For example, the presence of THC can be determined at very low concentrations in biological samples via SIM analysis of the trimethylsilyl derivative $C_{21}H_{29}O_2-Si(CH_3)_3$. An intense fragment ion for this compound at $m/z$ 371, resulting from loss of methyl radical from the molecular ion, has an empirical formula of $C_{23}H_{35}O_2Si$. The actual output of the electron multiplier as the mass analyzer scans over this mass is a curve having a maximum value at about $m/z$ 371.25 (Figure 2.4; see Section 1.5.2), with the peak width determined by the resolution of the instrument. For $\Delta M \sim 1$, setting the instrument to monitor $m/z$ 371.0 will result in less sensitivity than if it is set to

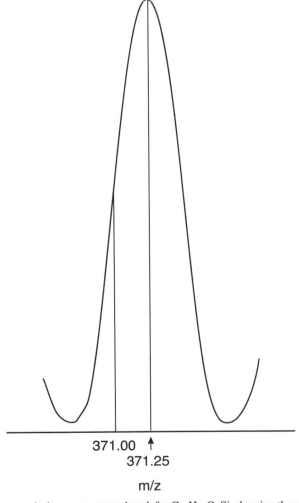

371.00 ↑
371.25

m/z

**Figure 2.4.** Low-resolution mass spectral peak for $C_{23}H_{35}O_2Si$, showing the effect of the mass defect on maximum intensity.

monitor $m/z$ 371.25, simply because $m/z$ 371.0 is not located at the top of the peak. The amount of sensitivity lost will depend on the resolution of the instrument—the greater the resolution (and the narrower the peak), the greater the loss in sensitivity.

Probably the most important application of mass defects is in very high resolution mass spectrometry (Section 1.3.4.4), which can provide exact mass measurements out to three or four decimal places. Consider two ions having the same nominal mass but different empirical formulas—for example, $^{12}C_{15}{}^{1}H_{10}{}^{14}N^{16}O_3{}^{35}Cl$ and $^{12}C_{14}{}^{1}H_8{}^{14}N_2{}^{16}O_3{}^{35}Cl$, both of which have a mass of 287. Using the values from Table 2.1, the mass defect for the first ion is

$$\Delta = 0.0000(15C) + 0.0078(10H) + 0.0031(1N) + (-0.0051)(3O) + (-0.0311)(1Cl)$$
$$= 0.0000 + 0.0780 + 0.0031 - 0.0153 - 0.0311 = 0.0347$$

while that of the second is

$$\Delta = 0.0000(14C) + 0.0078(8H) + 0.0031(2N) + (-0.0051)(3O) + (-0.0311)(1Cl)$$
$$= 0.0000 + .0604 + 0.0062 - 0.0153 - 0.0311 = 0.0202$$

It is easy to see that these two ions, which differ slightly in their empirical formulas, can be distinguished from one another at high enough mass resolution. Conversely, because the empirical formulas of most organic compounds have mass defects that differ from one another in the third or fourth decimal place, high-resolution mass spectrometry can usually determine a unique empirical formula for an ion. This is powerful information, providing us with direct access either to the empirical formula for the compound (if the ion is the molecular ion) or to what fragments have been lost from the molecular ion to give the observed ion. Such information is helpful in determining fragmentation pathways and thus for developing a basis for understanding why molecules fragment the way they do (an example is given in Section 8.3).

## 2.2. CALCULATING PEAK INTENSITIES FROM ISOTOPIC ABUNDANCES

Two of the elements listed in Table 2.1 (hydrogen and fluorine) occur in nature almost exclusively as a single isotope. The small amount of deuterium ($^2H$) that does occur naturally (0.01%) can safely be ignored in mass spectrometry because its contribution is below the normal limits of detection, usually set at 0.1 to 0.5% of the base peak. Ions containing only hydrogen and fluorine will exhibit peaks at only one $m/z$ value because there are no other isotopes of either element to consider. Hydrogen fluoride, for example, gives a molecular ion at $m/z$ 20, and essentially no peak at $m/z$ 21. For these elements and for compounds containing only one atom of an element having a second naturally occurring isotope, isotopic abundance considerations are fairly trivial. Unfortunately, only a handful of compounds of interest to analytical organic chemists fall into this category, so that an understanding of the effects of isotopic abundances on peak intensities is important.

### 2.2.1. One or More Atoms of a Single Element

***2.2.1.1. Chlorine and Bromine.*** Figure 2.5 shows the spectra of three simple compounds containing chlorine and bromine. The most striking aspect of these spectra is the presence of intense peaks separated by 2 daltons. In the case of methyl bromide ($CH_3Br$), the pattern is particularly striking because both of the abundant ion clusters at high mass are doublets, with both peaks in each cluster having approximately equal intensity. On the other hand, the peaks at $m/z$ 36 and 38 in the spectrum of HCl and those at $m/z$ 61 and 63 in the spectrum of 1,2-dichloroethylene have patterns that are similar to one another, with the lower mass ion of each pair being approximately three times the intensity of the higher mass ion.

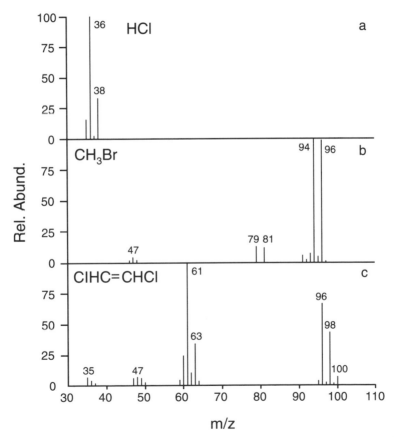

**Figure 2.5.** Mass spectra of three halogen-containing compounds: (*a*) hydrogen chloride, (*b*) methyl bromide, and (*c*) 1,2-dichloroethylene.

The data in Table 2.1 explain the sizes of the "doublet" peaks in the HCl and methyl bromide spectra. In the case of HCl, the molecular ion consists of two distinct isotopic entities: $H^{35}Cl^+$ ($m/z$ 36) and $H^{37}Cl^+$ ($m/z$ 38) in an approximate ratio of 75 to 25% (3:1), reflecting their relative natural abundances. Similarly, the molecular ion of methyl bromide consists of $CH_3^{79}Br^+$ ($m/z$ 94) and $CH_3^{81}Br^+$ ($m/z$ 96) in an approximate ratio of 50 to 50% (1:1). By further analogy, the ions at $m/z$ 35 and 37 in the HCl spectrum are in a 3:1 ratio reflecting the presence of $Cl^+$, and the ions of equal intensity at $m/z$ 79 and 81 in the spectrum of $CH_3Br$ reflect the presence of $Br^+$.

The ions at $m/z$ 61 and 63 in Figure 2.5c, in an approximate ratio of 3:1, also indicate the presence of one chlorine. Coupling this with the observation that $m/z$ 61 is exactly 35 daltons less than the molecular ion at $m/z$ 96, we can postulate by inference that the ion at $m/z$ 96 must be $C_2H_2^{35}Cl_2^+$, which also has $^{37}Cl$ components at $m/z$ 98 and 100. The relative sizes of the peaks at $m/z$ 96, 98, and 100 in the 1,2-dichloroethylene, however, are not so intuitive. In order to appreciate where these spectrum peaks come from, let us explore the process of determining isotopic abundance patterns in halogenated molecules.

The relative sizes of peaks due to natural isotopic abundances in cases involving a single chlorine or a single bromine can be put into a form that is useful for more generalized calculations. Using the notation $P(X)$ to denote the probability of a given isotope or set of isotopes occurring, the presence of one chlorine is reflected in the probabilities of the individual isotopes—that is,

$$P(^{35}Cl) = 0.75 \quad \text{and} \quad P(^{37}Cl) = 0.25$$

where the probabilities are simply the *approximate* natural isotopic abundances of $^{35}Cl$ and $^{37}Cl$, adjusted so that their sum is 1.0, rather than 100%. (Throughout the subsequent discussion the approximate natural abundances for chlorine and bromine will be used in order to emphasize mathematical relationships between the probabilities. The values in Table 2.1 should be used for more precise calculations.) The relative sizes of the two peaks generated by the presence of this chlorine are then given by the equation

$$[M^+]/[(M + 2)^+] = P(^{35}Cl)/P(^{37}Cl) = 0.75/0.25 = 100/33 = 3/1$$

where $M^+$ is the lower mass ion and $(M + 2)^+$ is the ion 2 daltons higher in mass. The situation is similar if we have one bromine in a molecule:

$$P(^{79}Br) = 0.5 \qquad P(^{81}Br) = 0.5$$

and

$$[M^+]/[(M + 2)^+] = P(^{79}Br)/P(^{81}Br) = 0.5/0.5 = 100/100 = 1/1$$

But consider now a situation in which two bromines occur in the same molecule. Then three combinations of isotopes are possible. In this case all of the combinations have approximately the same probability of occurrence, since the relative abundances of $^{79}Br$ and $^{81}Br$ are approximately equal:

$$P(2\ ^{79}Br) = (0.5)(0.5) = 0.25 \qquad P(^{79}Br)(^{81}Br) = (0.5)(0.5) = 0.25$$
$$P(2\ ^{81}Br) = (0.5)(0.5) = 0.25$$

Note that the probability for each combination is calculated by taking the product of the individual probabilities, since the probability of the second isotope being $^{79}Br$ or $^{81}Br$ is completely independent of the nature of the first.

For molecules having exactly one atom of each isotope present, there are two different orientations that are not identical but still indistinguishable by mass spectrometry—$^{79}Br^{81}Br$ and $^{81}Br^{79}Br$. The probability that *either one or the other* orientation will occur is equal to the sum of the individual probabilities for the two orientations. Since in this case the two probabilities are the same, the actual relative probabilities for the three peaks are

$$[M^+]/[(M + 2)^+]/[(M + 4)^+] = P(2\ ^{79}Br)/[P(^{79}Br)(^{81}Br) + P(^{81}Br)(^{79}Br)]$$
$$/P(2\ ^{81}Br)$$
$$= (0.25)/[(0.25) + (0.25)]/(0.25)$$
$$= (0.25)/(0.5)/(0.25)$$
$$= 50/100/50 = 1/2/1$$

Without taking these orientations into account, the sum of the probabilities for all isotopes is only $0.25 + 0.25 + 0.25 = 0.75$. When both orientations are considered, the total probability is 1.0.

Two chlorines in a molecule present a similar situation, although in this case the two isotopes of chlorine do not occur with equal probability:

$$P(2\ ^{35}Cl) = (0.75)^2 = 0.563 \quad P(^{35}Cl)(^{37}Cl) = (0.75)(0.25) = 0.188$$
$$P(2\ ^{37}Cl) = (0.25)^2 = 0.063$$

Again, the mass spectrometer cannot distinguish between $^{35}Cl^{37}Cl$ and $^{37}Cl^{35}Cl$, so that

$$[M^+]/[(M + 2)^+]/[(M + 4)^+] = P(2\ ^{35}Cl)/[P(^{35}Cl)(^{37}Cl) + P(^{37}Cl)(^{35}Cl)]$$
$$/P(2\ ^{37}Cl)$$
$$= (0.563)/[(0.188) + (0.188)]/0.063$$
$$= (0.563)/(0.375)/(0.063)$$
$$= 100/66/11$$

Careful examination of the ions at $m/z$ 96, 98, and 100 in the 1,2-dichloroethylene spectrum (Figure 2.5c) shows that they have these approximate intensities relative to one another, confirming the presence of two chlorine atoms in the molecular ion.

The presence of three bromines in the same molecule produces a slightly more complex situation that is approachable as an extension of the ideas developed so far:

$P(3\ ^{79}Br) = (0.5)^3 = 0.125$     $P(2\ ^{79}Br)(^{81}Br) = (0.5)^2(0.5) = 0.125$
$P(^{79}Br)(2\ ^{81}Br) = (0.5)(0.5)^2 = 0.125$     $P(3\ ^{81}Br) = (0.5)^3 = 0.125$

(It needs to be reemphasized that the relative probabilities for these four species are approximately equal only because the relative natural abundances of $^{79}Br$ and $^{81}Br$ are virtually the same.) For molecules containing two atoms of $^{79}Br$ and one of $^{81}Br$, three different orientations are not identical but also not distinguishable by mass spectrometry: $^{79}Br^{79}Br^{81}Br$, $^{79}Br^{81}Br^{79}Br$, and $^{81}Br^{79}Br^{79}Br$. Similarly, for molecules having a single $^{79}Br$ and two atoms of $^{81}Br$, three analogous orientations are possible. Therefore,

$[M^+]/[(M + 2)^+]/[(M + 4)^+]/[(M + 6)^+]$
$\qquad = P(3\ ^{79}Br)/3 \times P(2\ ^{79}Br)(^{81}Br)/3 \times P(^{79}Br)(2\ ^{81}Br)/P(3\ ^{81}Br)$
$\qquad = (0.125)/3(0.125)/3(0.125)/(0.125)$
$\qquad = 1/3/3/1$

Finally, with one chlorine and one bromine in a molecule,

$P(^{35}Cl)(^{79}Br) = (0.75)(0.5) = 0.375$     $P(^{37}Cl)(^{79}Br) = (0.25)(0.5) = 0.125$
$P(^{35}Cl)(^{81}Br) = (0.75)(0.5) = 0.375$     $P(^{37}Cl)(^{81}Br) = (0.25)(0.5) = 0.125$

In this case, the two middle mass orientations are not only not identical, but whether or not they are distinguished by mass spectrometry depends on the mass resolution of the spectrometer. The four isotopes have different mass defects so that the $^{35}Cl^{81}Br$ combination has a slightly different absolute mass from that of $^{37}Cl^{79}Br$ (115.885 vs. 115.884, to be precise). Under low-resolution conditions the two will not be distinguishable, and

$[M^+]/[(M + 2)^+]/[(M + 4)^+] = P(^{35}Cl)(^{79}Br)/[P(^{37}Cl)(^{79}Br) +$
$\qquad\qquad P(^{35}Cl)(^{81}Br)]/P(^{37}Cl)(^{81}Br)$
$\qquad\qquad = (0.375)/[(0.125) + (0.375)]/(0.125)$
$\qquad\qquad = (0.375)/(0.5)/(0.125) = 75/100/25$
$\qquad\qquad = 3/4/1$

At very high resolution, however, four peaks will be discernible, the middle two (at M + 2) separated by 0.001 dalton, in a ratio of (0.375):(0.125):(0.375):(0.125) or 3:1:3:1.

Calculating these intensities each time chlorine and bromine are encountered in a molecule would be a nuisance. Fortunately, many reference books on mass spectrometry contain the results of these calculations—some references including intensities for up to 8 or 10 halogen atoms. Figure 2.6 shows in graphic form the results we have just calculated, as well as those for 3 and 5 chlorines.

The patterns in Figure 2.6 have visual impact that will become apparent as more and more mass spectra are encountered. Each pattern is characteristic of a specific halogen content in the molecule, and with experience one can learn to identify the presence of various halogen combinations almost immediately upon glancing at a

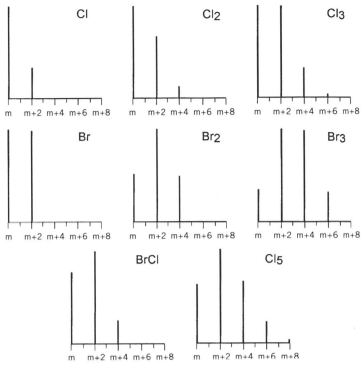

**Figure 2.6.** Intensities for some isotope clusters containing bromine and chlorine.

spectrum (Figure 2.7). Looking for chlorine and bromine is one of the first steps in mass spectral interpretation, simply because the evidence is almost always so obvious.

*Problem 2.3.* Using the same logic as for three bromines, calculate the relative intensities of the four peaks you would expect to see from the presence of three chlorines in a molecule. Remember to use the isotopic abundances for chlorine, not bromine!

One consequence of the unusually high natural abundances of $^{37}$Cl and $^{81}$Br is that, with ions containing more than three chlorines or one bromine, the lowest mass ion in the ion cluster is no longer the most abundant. In fact, for highly chlorinated and brominated compounds, the intensity of the lowest mass ion in the cluster may be rather small. This leads to a problem in nomenclature since we do not normally refer to ions having specific isotopic content. We might talk about a $C_5Cl_5^+$ ion, but we would rarely say that an ion was $^{12}C_5^{37}Cl_5^+$, simply because, for most mass spectral interpretation, the specific isotopes are not important. For uniformity, however, we will refer to an ion by the lowest mass peak in the ion cluster *regardless of whether or not it is the most intense ion in the cluster.*

The presence of more than one intense ion in an ion cluster means that care must be exercised when calculating losses from one ion cluster to another. For example, in

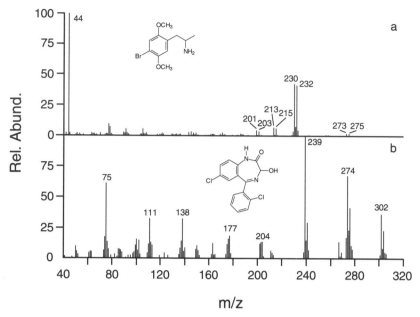

**Figure 2.7.** Mass spectra of 4-bromo-2,5-dimethoxyamphetamine and lorazepam showing the presence of halogen(s) in the high mass ion clusters.

the spectrum of methyl bromide (Figure 2.5*b*), the molecular ion, from the definition above, is *m/z* 94, corresponding to $^{12}CH_3{}^{79}Br^+$, even though the ion at *m/z* 96 is almost equally as intense. The spectrum also contains two ions of nearly equal intensity at *m/z* 79 and 81, which correspond to $^{79}Br^+$ and $^{81}Br^+$, respectively. Knowing this, it makes no sense to talk about a loss from *m/z* 94 to 81 (a loss of 13 daltons) because these two ions contain different isotopes of bromine. A nuclear transformation would have to occur for this loss to take place! Instead, loss of a methyl radical ($CH_3\cdot$) from *m/z* 94 accounts for the ion at *m/z* 79, just as loss of methyl radical from *m/z* 96 accounts for the ion at *m/z* 81, since both the parent and fragment ions must contain the same isotope of bromine.

### 2.2.1.2. Statistical Factors with Multiple Numbers of Atoms.

For ions containing several atoms of the same element, the probability calculations in the previous section were seen to depend not only on the natural abundances of the individual isotopes but also upon the number of indistinguishable orientations of the isotopes. Values for these "statistical factors" are seen most clearly in intensities for ion clusters containing several bromines (Figure 2.6): when one bromine is present, the values are 1:1 (i.e., there is only one possible orientation for the isotopes in each of the two ions); when there are two bromines, the values are 1:2:1 (one orientation for the lowest and highest mass ions and two for the ion containing one atom of each isotope); for three bromines, they are 1:3:3:1.

These numbers are also the coefficients obtained in the binomial expansion of the expression $(a + b)^n$, where $n$ is an integer. When $n = 1$, we have $(1)a + (1)b$; when $n = 2$, $(1)a^2 + 2ab + (1)b^2$; when $n = 3$, $(1)a^3 + 3a^2b + 3ab^2 + (1)b^3$, and so forth. Calculating these coefficients by brute force for higher values of $n$ is tedious, but fortunately they form a pattern called *Pascal's triangle* (Figure 2.8), named after the seventeenth-century French mathematician. In this array, the values of the binomial expansion coefficients are given horizontally in the same row as the value of $n$ shown in the left-hand column.

Three patterns emerge from this array that are useful in this discussion. First, the coefficients on the left edge of the triangle (the left-most diagonal column of Figure 2.8) are all 1. The coefficient in this diagonal column for any row is the statistical factor for the lowest mass peak in the isotope cluster. Since all of the isotopes in this ion are the same (the lowest mass isotope), all arrangements of the atoms in this ion are identical, regardless of how many atoms of the particular element there are, and there is only one possible orientation—therefore, the "statistical factor" is 1.

Second, the coefficients in the diagonal column second from the left in Figure 2.8 all have the same value as $n$ for that horizontal row. This coefficient helps determine the relative size of the ion that is next highest in mass. For chlorine and bromine, this corresponds to an ion 2 daltons higher in mass than the lowest mass peak [which we will denote as $(M + 2)^+$ if the ion is the molecular ion, and $(P + 2)^+$ for other ions in the spectrum]; the same is true for oxygen and for the two most abundant isotopes of sulfur (Table 2.1). For carbon and nitrogen, this ion is only 1 dalton higher [which we will denote as $(M + 1)^+$ or $(P + 1)^+$]. The value for this coefficient is always the number of atoms of the element under consideration. Mathematically, this corresponds to the number of distinguishable ways in which you can arrange $n$ objects when one of the objects differs from the other $(n - 1)$, called the "combination of $n$ things taken one at a time." For an ion having five chlorines, the peak at

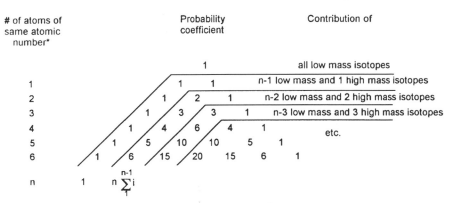

**Figure 2.8.** Binomial expansion (Pascal's) triangle. $\sum_{1}^{n-1} i = n(n-1)/2$. (*Assume the element in question has only two isotopes contributing to isotope peak ratios.)

$(M + 2)^+$ will contain exactly four $^{35}$Cls and one $^{37}$Cl. There are five different ways of writing this combination that cannot be distinguished by mass spectrometry: $^{35}$Cl$^{35}$Cl$^{35}$Cl$^{35}$Cl$^{37}$Cl, $^{35}$Cl$^{35}$Cl$^{35}$Cl$^{37}$Cl$^{35}$Cl, $^{35}$Cl$^{35}$Cl$^{37}$Cl$^{35}$Cl$^{35}$Cl, $^{35}$Cl$^{37}$Cl$^{35}$Cl$^{35}$Cl$^{35}$Cl, and $^{37}$Cl$^{35}$Cl$^{35}$Cl$^{35}$Cl$^{35}$Cl. Therefore, the "statistical factor" is 5.

A third pattern that results from the algebraic derivation of the elements of Pascal's triangle (Figure 2.8) is even more interesting. Every number in the array is equal to the sum of the two numbers that flank it in the row immediately above. Thus, the third coefficient in the row where $n = 6$, which has the value 15, is obtained by adding the two numbers immediately above it in the triangle, namely 10 and 5. Similarly, the next coefficient (20) is the sum of the two 10s that appear just above it in the array. On the basis of this one relationship, the entire array can be derived.

This pattern allows us to calculate a generalized formula for a coefficient in the third diagonal column from the left of Figure 2.8. Consider the value of this coefficient in the row where $n = 5$. Its value (10) is the sum of the two numbers above it (4 and 6). At the same time the 6 is also in the third diagonal column from the left in the row where $n = 4$, and so is the sum of the two numbers above it (3 and 3). The second of those numbers is also in the third diagonal column from the left in the row where $n = 3$, and so is equal to the two numbers above it (2 and 1). Putting all of this information together, we have $10 = 4 + [3 + (2 + 1)]$, which is simply the sum of all of the integers having a value less than $n$ (in this case, 5). You can quickly convince yourself that this is a general formula by noting that the third coefficient from the left in the row where $n = 6$ is 15, which is equal to 5 plus the same 10 whose value we just derived. This formula, generalized for any value of $n$, is written as

$$\sum_{1}^{n-1} i = (n - 1) + (n - 2) + \ldots + 3 + 2 + 1$$

where the notation on the left stands for the summation (the capital Greek letter sigma) of all of the integers from 1 to $(n - 1)$.

Although these coefficients can be calculated for any value of $n$, the summation notation still is not convenient for substituting into equations. To develop a useful form, consider the situation where four bromines occur in a molecule. The ion third lowest in mass [the $(M + 4)^+$ ion] will contain exactly two atoms each of $^{79}$Br and $^{81}$Br. In order to determine how many different arrangements of these atoms are possible, we need to know the number of ways in which $n$ objects can be arranged when two of those objects are different from the other $(n - 2)$—in other words, the combination of $n$ things taken two at a time. Mathematically, the formula for calculating this statistical factor for "the combination of $n$ things taken $m$ at a time," where $m$ is an integer less than $n$, is

$$C(n, m) = \frac{n(n - 1) \ldots (n - m + 1)}{m!}$$

where $C(n, m)$ denotes the combination of $n$ things taken $m$ at a time, the numerator on the right side of the equation is the product of all the integers from $n$ down to

$(n - m + 1)$, and the denominator $m!$ ($m$ factorial) is the product of all integers from 1 through $m$.

For the specific situation where $n$ things are taken 2 at a time, the expression above becomes

$$C(n, 2) = \frac{n(n-1)}{2!}$$

Since this coefficient (statistical factor) is the same as that determined above by summation, we now have an expression for the sum that can be used in generalized equations, namely

$$\sum_{1}^{n-1} i = C(n, 2) = \frac{n(n-1)}{2} \tag{2.1}$$

Applying this equation to the $(M + 4)^+$ peak for 4 bromines, we have

$$C(4, 2) = \frac{(4)(3)}{2!} = \frac{(4)(3)}{(1)(2)} = 6$$

This is exactly the number of indistinguishable, but not identical, arrangements that can be written for the bromine isotopes in this ion: $^{79}Br^{79}Br^{81}Br^{81}Br$, $^{81}Br^{81}Br^{79}Br^{79}Br$, $^{79}Br^{81}Br^{79}Br^{81}Br$, $^{81}Br^{79}Br^{81}Br^{79}Br$, $^{79}Br^{81}Br^{81}Br^{79}Br$, and $^{81}Br^{79}Br^{79}Br^{81}Br$.

### 2.2.1.3. Isotope Ratios for Carbon-Containing Compounds—The (M + 1)+ Ion.

We have discussed peak intensities for the halogens in some detail in order to develop an empirical feeling for how peak intensities in isotopic clusters depend upon both the natural abundances of the isotopes and the statistical factors related to the number of atoms of the element present in the molecule. This same logic can now be applied to calculating the relative sizes of the $(M + 1)^+$ and $M^+$ peaks for a compound containing $n$ carbon atoms. For the time being, assume that no other atoms contribute to the size of the $(M + 1)^+$ peak.

Following the symbols and the logic developed in the previous sections, the probability determining the size of the molecular ion $M^+$ is that of finding $n$ $^{12}C$ atoms in the same molecule:

$$P(M) = P(n\ ^{12}C) = (0.989)^n$$

which is simply the probability of finding one $^{12}C$ (0.989, its natural abundance from Table 2.1) multiplied $n$ times. Similarly, the probability determining the size of the $(M + 1)^+$ peak is that of finding a molecule with exactly $(n-1)$ $^{12}C$ atoms and one $^{13}C$. This is expressed by

$$P(M + 1) = P[(n - 1)\ ^{12}C + 1\ ^{13}C] = nP[(n - 1)\ ^{12}C)] \times P(^{13}C) = $$
$$n(0.989)^{n-1}(0.011)$$

where the statistical coefficient $n$ arises because there are $n$ different positions for the $^{13}C$ atom that are indistinguishable by mass spectrometry (Section 2.2.1.2; see also Figure 2.8). The 0.011 probability comes from the natural abundance of $^{13}C$.

The relative intensities of the $M^+$ and $(M + 1)^+$ peaks are calculated by the ratio of their probabilities:

$$[(M + 1)^+]/[M^+] = P(M + 1)/P(M)$$
$$= n(0.989)^{n-1}(0.011)/(0.989)^n$$
$$= n(0.011)/(0.989)$$
$$= n(0.0111)$$

To convert this equation to percentages, both numerator and denominator are multiplied by 100, giving

$$n(0.011)(100)/(100) = n \times 1.1/100$$
$$= n \times 1.1\% \qquad (2.2)$$

In retrospect, we can see that it is somewhat fortuitous that the 1.1% figure appearing in Eq. (2.2) just happens to correspond to the natural isotopic abundance of $^{13}C$. It is only because the natural abundance of $^{12}C$ is so close to 100% that its value essentially falls out of the equation.

Equation (2.2) has several consequences that may not be immediately apparent. First, if the number of carbon atoms in a molecule is known, their contribution to the size of the $(M + 1)^+$ peak can be calculated. For example, since the molecular formula for $\Delta^9$-tetrahydrocannabinol (THC) is $C_{21}H_{30}O_2$, and since neither hydrogen or oxygen have isotopes that contribute to the $(M + 1)^+$ ion, the size of the $m/z$ 315 ion relative to the molecular ion at $m/z$ 314 in the spectrum of THC should be

$$n \times 1.1\% = 21 \times 1.1\% = 23.1\%$$

When this number is compared with that actually measured for this compound in Table 1.3 (where this ratio is 16.8/70.5 = 23.8%), we see that the observed intensity is close, but not identical, to the calculated value. This reflects an experimental error inherent in mass spectrometry, which is that the relative abundances of larger peaks are reproducible under normal circumstances only to about ±10%. For peaks under about 5% relative abundance, the relative error is often even higher. Using these criteria, the agreement seen here is actually quite good.

This type of calculation may provide other important clues about various ions in the spectrum. Toluene (methylbenzene; $C_7H_8$), for example, exhibits two intense high mass ions at $m/z$ 91 and 92 with relative intensities of 100 and 76%, respectively. It is tempting at first to assume that $m/z$ 91 is the molecular ion, since it is the most abundant ion in the highest mass cluster in the spectrum. However, assuming that this compound contains only carbon, hydrogen, nitrogen, and perhaps oxygen (there are no indications of any unusual isotopic patterns; see Figure 4.6b), we can place an upper limit on the relative size of the $(M + 1)^+$ ion for this compound. Since an organic compound having a molecular weight around 90 daltons *cannot* contain more than 7 carbon atoms ($7 \times 12 = 84$), and since hydrogen, nitrogen, and oxygen do not contribute substantially to the size of the $(M + 1)^+$ ion, then the maximum size expected for the $m/z$ 92 ion (assuming $m/z$ 91 is the molecular ion) is about $7 \times 1.1\% = 7.7\%$. Since the actual size of $m/z$ 92 is 76%, far in excess of that expected, we must conclude that $m/z$ 91 is not the molecular ion. A more likely explanation (borne out by the size of the $m/z$ 93 ion, which is about 6%) is that $m/z$ 92 is the molecular ion and $m/z$ 91 arises from facile loss of hydrogen radical from the molecular ion. We will return for a more detailed look at the molecular ion region of the toluene spectrum in Section 2.2.1.5.

Reversing this process, the size of the $(M + 1)^+$ peak can be used to calculate the number of carbon atoms in the ion. Consider, for instance, a spectrum that shows a molecular ion at $m/z$ 118 and an ion at $m/z$ 119 having an intensity of approximately 9% that of the $m/z$ 118 ion. If the compound contains no heteroatoms other than oxygen, we can solve for the number of carbon atoms in this compound as follows:

$$9.0\% \approx n \times 1.1\% \quad \text{or} \quad n \approx 9.0\%/1.1\% \approx 8 \text{ carbon atoms}$$

In real-life situations, this formula usually yields less information than we might like. Most importantly, it works *only* if there are no interferences from other ions. This means that its use for fragment ions is always risky, as we just saw with toluene. In addition the 10% experimental error inherent in measuring relative ion intensities becomes limiting for higher numbers of carbons. For example, an $(M + 1)^+$ ion having a relative intensity of 22.5% has an experimentally usable relative intensity of $22.5 \pm 2.3\%$, which covers the range from 20.2 to 24.8%. If we know the relative intensity is exactly 22.5%, we can conclude that there are 20 ($- 22.5\%/1.1\%$) carbon atoms in the molecule, but with experimental error this number is actually $20 \pm 2$ carbons. This is not very helpful for determining the exact molecular formula for an unknown. Nonetheless, for smaller molecules Eq. 2.2 can prove useful, as you will see in solving the problems in this book.

The logic used to calculate the relative intensity of the $(M + 1)^+$ ion for carbon is also applicable to $^{15}N$ and $^{33}S$ for their contributions to the $(M + 1)^+$ ion and to $^{18}O$ and $^{34}S$ for their contributions to the $(M + 2)^+$ ion. The results of the calculations for these elements are as follows:

For $^{15}N$:    $[(M + 1)^+]/[M^+] = n(0.0036)/(0.9964) = n \times 0.0036 \Rightarrow n \times 0.36\%.$

For $^{33}S$:    $[(M + 1)^+]/[M^+] = n(0.0076)/(0.9503) = n \times 0.0080 \Rightarrow n \times 0.80\%.$

For $^{18}O$:    $[(M + 2)^+]/[M^+] = n(0.0020)/(0.9976) = n \times 0.0020 \Rightarrow n \times 0.20\%.$

For $^{34}S$:    $[(M + 2)^+]/[M^+] = n(0.0420)/(0.9503) = n \times 0.0442 \Rightarrow n \times 4.42\%.$

Because the abundance of $^{32}S$ is substantially different from 100%, the percentages by which $n$ must be multiplied for both $^{33}S$ and $^{34}S$ are somewhat different from the actual abundances of these isotopes.

The effect on the size of the $(M + 1)^+$ ion when both carbon and nitrogen are present in the molecule is additive, since both $^{13}C$ and $^{15}N$ cannot contribute to this peak at the same time [in that case the mass of the ion would be (M + 2)]. That this should be so can be seen from the fact that an ion containing a certain number of $^{12}C$'s and $^{14}N$'s and one $^{13}C$ will have a different exact mass from an ion having $^{12}C$'s, $^{14}N$'s, and one $^{15}N$, and thus will be distinguishable at high resolution. For an ion containing both carbon and nitrogen, then, the relative size of the $(M + 1)^+$ ion at low resolution is given by

$$[(M + 1)^+]/[M^+] = \text{(no. of carbons} \times 1.1\%) + \\ \text{(no. of nitrogens} \times 0.36\%) \tag{2.3}$$

Similarly, the intensity of the $(M + 1)^+$ ion for an ion containing carbon, nitrogen, and sulfur would be determined by the additive effects of isotopes from all three elements.

### 2.2.1.4. Isotope Ratios for Carbon-Containing Compounds—The (M + 2)+ Ion.

Compounds containing large numbers of carbon atoms also show a detectable effect on the size of the $(M + 2)^+$ ion due to the contributions from two $^{13}C$'s. Calculating this effect, however, is not straightforward. It is tempting to present Eq. (2.4) with little comment, but because its form is not intuitive, the following discussion of its derivation is presented.

To calculate the intensity of the $(M + 2)^+$ ion relative to that of the $M^+$ ion for a compound containing $n$ carbon atoms, we must again assume that there are no other atoms that contribute to the size of the $(M + 2)^+$ peak. As we saw in the previous section, the probability determining the size of $M^+$ is $P(M) = P(n^{12}C) = (0.989)^n$. The size of the $(M + 2)^+$ peak is now determined by the probability of finding exactly $n - 2$ $^{12}C$ atoms and two $^{13}C$ atoms in the same molecule. This is expressed by

$$P(M+2) = P[(n-2)^{12}C + 2^{13}C] = (\sum_{i=1}^{n-1} i)(0.989)^{n-2}(0.011)^2$$

in which the statistical coefficient is obtained from the third diagonal column from the left in the $n$th horizontal row of Figure 2.8. The relative sizes of the two peaks is simply the ratio of the two probabilities:

$$\frac{P(M+2)}{P(M)} = (\sum_{i=1}^{n-1} i)\frac{(0.989)^{n-2}(0.011)^2}{(0.989)^n}$$

$$= (\sum_{i=1}^{n-1} i)\frac{(0.011)^2}{(0.989)^2}$$

We have already shown [Eq. (2.1)] that

$$\sum_{i=1}^{n-1} i = C(n,2) = \frac{n(n-1)}{2}$$

so that

$$\frac{P(M+2)}{P(M)} = \frac{n(n-1)(0.011)^2}{2(0.989)^2} = \frac{n^2(0.011)^2}{2(0.989)^2} - \frac{n(0.011)^2}{2(0.989)^2}$$

This equation is too complex to use on a routine basis so, as a rough approximation, we assume that $(0.989)^2 \sim 1$ and $n^2 >> n$ if $n$ is relatively large (and if $n$ isn't large, the size of this peak is going to be very small anyway). Making these approximations, the second term of the equation above can be ignored because it is very small compared to the first term, and the denominator of the remaining term becomes $2 \times 1 = 2$, or

$$\frac{[(M+2)^+]}{[M^+]} = \frac{n^2(0.011)^2}{2}$$

To translate this into percentage terms, we multiply numerator and denominator by $(100)^2$, so that

$$\frac{[(M+2)^+]}{[M^+]} = \frac{n^2(0.011)^2(100)^2}{(2)(100)(100)} = \frac{n^2(1.1)^2}{(200)(100)} = \frac{(n \times 1.1)^2\%}{200} \tag{2.4}$$

This is the form in which this equation is usually found in the mass spectral literature.

The $^{13}C$ contribution to the size of the $(M + 2)^+$ ion for compounds containing few carbon atoms is small—even the presence of 10 carbons leads to a peak of only about 0.6% relative to the intensity of the $M^+$ ion and may be visible only if the $M^+$ peak itself is at least moderately intense. For larger numbers of carbons, contributions increase rapidly. Again using the example of THC, which has 21 carbons, the contribution of two $^{13}C$'s to the $(M + 2)^+$ ion is $(21 \times 1.1)^2/200 = (23.1)^2/200 = 533.6/200 = 2.67\%$, a peak that can hardly be ignored.

Of course $^{13}C$ is not the only contributor to the $(M + 2)^+$ ion. In addition to the obvious contributions of $^{37}Cl$ and $^{81}Br$, $^{34}S$ makes a substantial contribution, and $^{18}O$ makes a small, but detectable contribution, especially if more than one oxygen is present. If either oxygen or sulfur are present together with carbon in the same molecule, the effects are additive and independent of one other, for reasons discussed in the previous section. For organic compounds having no heteroatoms other than nitrogen and oxygen, the relative size of the $(M + 2)^+$ ion is commonly given by the following equation:

$$[(M + 2)^+]/[M^+] = \text{(no. of carbons} \times 1.1)^2\%/200 + \text{(no. of oxygens} \times 0.20\%) \tag{2.5}$$

The calculated size of the $(M + 2)^+$ ion for THC $(C_{21}H_{30}O_2)$, taking into account all contributors, is thus given by

$$[(M + 2)^+]/[M^+] = (21 \times 1.1)^2\%/200 + (2 \times 0.2\%) = 2.67 + 0.4 = 3.1\%$$

which compares with the observed value of 3.5% (Table 1.3).

***Problem 2.4.*** Buckminsterfullerene, known familiarly as "buckyball," is a recently synthesized form of carbon having the shape of a soccer ball. This compound was named after Buckminster Fuller, the inventor of the geodesic dome, which has a similar shape. The molecular formula for buckminsterfullerene is $C_{60}$ (there are no hydrogens!!), and the base peak in its mass spectrum is the molecular ion. Calculate the intensities of the $(M + 1)^+$ and $(M + 2)^+$ ions for this compound.

### 2.2.1.5. Overlapping Ion Clusters—Contributions from $^{13}C$ Only. The
molecular ion region in the mass spectrum of the aromatic hydrocarbon toluene $(C_7H_8)$ is illustrated in expanded form in Figure 2.9 (the full spectrum of toluene is shown in Figure 4.6*b*). The observed intensities for these ions are given at the right side of Table 2.2. The ion at *m/z* 91 is not the molecular ion (Section 2.2.1.3), but let us now look more closely and see how the isotopic contributions from each ion in the cluster influences the intensities of subsequent ions.

This cluster of ions begins at *m/z* 89 ($^{12}C_7H_5{}^+$) with an ion having an observed intensity of 3.8%. The size of this peak is completely independent of isotopic abundance information—the next lower mass ion occurs at *m/z* 78. However, because there is a peak at *m/z* 89 for $^{12}C_7H_5{}^+$, we now know that there *must* be a corresponding ion at *m/z* 90 for $^{13}C^{12}C_6H_5{}^+$ and that the size of this ion must be 7.7% (= 7 carbons × 1.1%) of the size of the ion at *m/z* 89. Since *m/z* 89 has an intensity of 3.8%, the ion at *m/z* 90 for $^{13}C^{12}C_6H_5{}^+$ must have an intensity of 3.8% × 7.7% = 0.3%, which is small, but measurable.

In this spectrum the actual relative abundance of *m/z* 90 is 9.9%, far greater than that predicted for $^{13}C^{12}C_6H_5{}^+$. Indeed it should not be surprising that there is a fragment ion $^{12}C_7H_6{}^+$ at *m/z* 90. On the other hand, we have just seen that $^{12}C_7H_6{}^+$ does not account for all of the observed intensity of 9.9%—only for this value *minus* the 0.3% contribution from $^{13}C^{12}C_6H_5{}^+$. It is worth repeating that these two ions at *m/z* 90 are indistinguishable only at low resolution. A high-resolution mass spectrometer would show them as separate ions since they have different absolute masses due to the different mass defects of $^{12}C$ and $^{13}C$. This also means that the contributions of these two ions to *m/z* 90 are additive at low mass resolution.

Since the observed intensity for $^{12}C_7H_6{}^+$ at *m/z* 90 has been shown to be 9.9 − 0.3 = 9.6%, there must be a corresponding ion at *m/z* 91 due to $^{13}C^{12}C_6H_6{}^+$ having an intensity of 7.7% of the abundance of the $^{12}C_7H_6{}^+$ ion at *m/z* 90 (there are still 7 carbons in this ion), which is 9.6 × 7.7 = 0.7%. In addition, because the $^{12}C_7H_6{}^+$ ion at *m/z* 90 is of moderate intensity, the possible contribution of $^{13}C_2{}^{12}C_5H_6{}^+$ to the *m/z* 92 ion must be considered. Using Eq. (2.4), this contribution is calculated to be $(7 × 1.1)^2\%/200$ of the size of the $^{12}C_7H_6{}^+$ at *m/z* 90, or 0.30 × 9.6 = 0.03%. Peaks of this size are ignored under normal conditions since they usually are at or below the level of background from the instrument.

Given a contribution of 0.7% to *m/z* 91 by $^{13}C^{12}C_6H_6{}^+$, the remaining 99.3% of this peak must be due to $^{12}C_7H_7{}^+$. As with the other ions just considered, the presence of the $^{12}C_7H_7{}^+$ ion at *m/z* 91 having an intensity of 99.3% dictates that there must be an ion at *m/z* 92 corresponding to $^{13}C^{12}C_6H_7{}^+$ and having an intensity of 7.7 × 99.3 = 7.6%. Also, since the $^{12}C_7H_7{}^+$ ion is so intense, we now expect a contri-

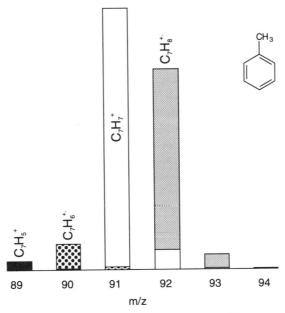

**Figure 2.9.** Overlapping ion clusters with contributions from $^{13}C$ only; the molecular ion region of toluene.

**Table 2.2. Overlapping Ion Clusters in Toluene (Contributions from $^{13}C$ Only)**

| $m/z$ | $C_7H_5^+$ | | $C_7H_6^+$ | | $C_7H_7^+$ | | $C_7H_8^+$ | Observed Intensity |
|---|---|---|---|---|---|---|---|---|
| 89 | 3.8° | | | | | | | = 3.8 |
| 90 | 0.3ᶜ | + | 9.6° | | | | | = 9.9 |
| 91 | | | 0.7ᶜ | + | 99.3° | | | = 100.0 |
| 92 | | | (0.03)ᶜ | + | 7.6ᶜ | + | 68.7° | = 76.3 |
| 93 | | | | | 0.3ᶜ | + | 5.2ᶜ | = (6.1) |
| 94 | | | | | | | 0.2ᶜ | = (<0.5) |

°Observed intensity.
ᶜCalculated intensity.

bution to the $m/z$ 93 ion from $^{13}C_2{}^{12}C_5H_7^+$, which turns out to be $(7.7)^2\%/200 \times 99.3\% = 0.3\%$, just above the limit of significance.

The molecular ion for toluene at $m/z$ 92 already has a 7.6% contribution from $^{13}C^{12}C_6H_7^+$, and a miniscule (and ignorable) contribution comes from $^{13}C_2{}^{12}C_5H_6^+$. Since the observed intensity of $m/z$ 92 is 76.3%, the contribution from $^{12}C_7H_8^+$ must be $76.3 - 7.6 = 68.7\%$, and the contribution of $^{13}C^{12}C_6H_8^+$ to $m/z$ 93 is calculated to be 7.7% of 68.7%, or 5.2%.

Now the total *calculated* contributions to $m/z$ 93 are 0.3% from $^{13}C_2{}^{12}C_5H_7^+$ and 5.2% from $^{13}C^{12}C_6H_8^+$—a total of 5.5%. The *observed* intensity of this ion in this spectrum is 6.1%, and there are no other ions that can reasonably account for the size of this peak. The difference between 5.5 and 6.1% is right at the 10% cutoff for experimental error in intensity measurement, so we can be satisfied that the observed and calculated intensities do, in fact, agree.

Finally, there should be a contribution of 68.7% $\times$ $(7.7)^2/200 = 0.2\%$ from $^{13}C_2{}^{12}C_5H_8{}^+$ to the ion at $m/z$ 94. All we know about the actual intensity for this ion is that it is less than 0.5%, in keeping with the calculated value.

***2.2.1.6. Silicon.*** Bromine and chlorine produce striking isotopic patterns that usually give immediate visual clues in the mass spectrum about their presence in a molecule. Silicon, with three isotopes having detectable abundances (Table 2.1), produces isotopic patterns that are more subtle but nonetheless distinctive. Figure 2.10 shows isotope clusters for ions containing one and four silicons, respectively. The ion at $m/z$ 73 is often encountered in the spectra of trimethylsilyl derivatives of compounds containing $-OH$ or $-NH_2$ groups, for example. Although the intensity of $m/z$ 74 is not particularly striking for an ion of that mass (a compound containing four or five carbon atoms would show a similar intensity), the intensity of the ion at $m/z$ 75 is too great to be due to contributions from $^{13}C$ alone.

The isotope cluster beginning at $m/z$ 281 more readily catches the eye. Ions such as this often occur when bleed from the chromatographic column or from a silicone septum becomes a problem. Other silicon-containing clusters from the same sources

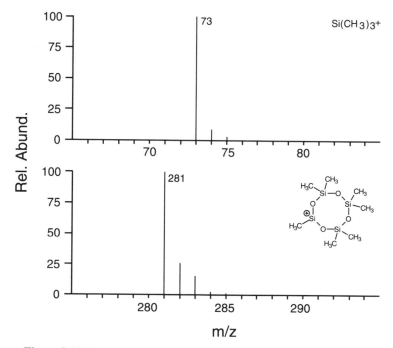

**Figure 2.10.** Isotope clusters for ions containing one and four silicons.

may be observed at *m/z* 207 and 355. As with the *m/z* 73 cluster, the $(P + 1)^+$ ion has an intensity that is typical for carbon-containing ions of similar mass, but the sizes of $(P + 2)^+$ and $(P + 3)^+$ ions are inconsistent with the presence of carbon alone. Learning the pattern for these silicon-containing ions will help identify their presence as background ions in the mass spectra of other compounds.

### 2.2.2. Complex Isotope Clusters

***2.2.2.1. Sulfur Dioxide.*** Most complex compounds analyzed by mass spectrometry contain carbon and at least one heteroatom. The presence of two or more elements having higher mass isotopes in the same molecule complicates the situation. A couple of specific examples should serve as illustration.

In the spectrum of sulfur dioxide (Figure 2.11), the $(M + 2)^+$ ion at *m/z* 66 is larger than the $(M + 1)^+$. These intensities are inconsistent with the presence of carbon alone, but the ion at *m/z* 66 is not intense enough to indicate chlorine A *somewhat* intensified $(M + 2)^+$ ion is rather uniquely typical of sulfur, although a compound containing a number of carbons and one or more sulfurs can be confused with an ion containing silicon under some circumstances.

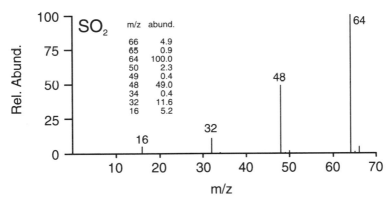

**Figure 2.11.** Isotopic abundances with contributions from two different elements (sulfur dioxide).

**Table 2.3. Calculated Relative Abundances of the Molecular Ion Cluster (Sulfur Dioxide)**

| | | |
|---|---|---|
| *m/z* 64 | $^{32}S^{16}O^{16}O$ | 100% |
| *m/z* 65 | $^{33}S^{16}O^{16}O \Rightarrow (0.76/95.0) \times 100\%$ (for 1 S) = | 0.8% |
| *m/z* 66 | $^{34}S^{16}O^{16}O \Rightarrow (4.2/95.0) \times 100\%$ (for 1 S) + | |
| | $^{32}S^{18}O^{16}O$     $2(0.2/99.8) \times 100\%$ (for 2 O) = 4.4 + 0.4 = | 4.8% |
| *m/z* 67 | $^{33}S^{18}O^{16}O \Rightarrow (0.76/95.0) \times 2(0.2/99.8) \times 100\%$ = | 0.0032% |
| *m/z* 68 | $^{34}S^{18}O^{16}O \Rightarrow (4.2/95.0) \times 2(0.2/99.8) \times 100\%$ + | |
| | $^{32}S^{18}O^{18}O$     $(0.2/99.8)^2 \times 100\% = 0.0177 + 0.0004 =$ | 0.018% |

Sulfur dioxide contains two elements each having more than one stable isotope—sulfur, as $^{32}S$, $^{33}S$, and $^{34}S$, and oxygen, as $^{16}O$ and $^{18}O$. We therefore expect a cluster of ions above the molecular ion at $m/z$ 64. Calculating the relative size of the ion at $m/z$ 65, as shown in Table 2.3, follows from previous discussions, since it involves only the contribution of $^{33}S^{16}O_2$. At low mass resolution, the ion at $m/z$ 66 has contributions from two different species: $^{34}S^{16}O_2$ and $^{32}S^{16}O^{18}O$. This situation has precedent—because the two ions have different absolute masses (the mass defects of sulfur and oxygen being different), their contributions at low resolution are calculated independently and are then added together [see Eq. (2.3)].

For the very low intensity ions at $m/z$ 67 and 68, however, a different situation applies. In these cases, both elements contribute higher mass isotopes to the ion at the same time. For each contributing ion, the overall probability is calculated by multiplying the probabilities of the individual isotopic abundances together. Here, to produce an ion at $m/z$ 67, the presence of $^{33}S$ demands the additional presence of exactly one $^{18}O$; we cannot have both $^{33}S$ and $^{34}S$ together since there is only one sulfur! Consider a more familiar example: The probability of rolling "snake eyes" (two 1's) using two dice is calculated by multiplying the probability of rolling one 1 (which is $\frac{1}{6}$, assuming that each side of the die may be rolled with equal facility) by the probability of obtaining a second 1—that is, $\frac{1}{6} \times \frac{1}{6} = \frac{1}{36}$. The relative size of $m/z$ 67 in the sulfur dioxide spectrum, then, is found by multiplying the relative probability of finding $^{33}S$ (0.76/95.0) by the relative probability of finding one $^{18}O$ (which, in this case, is $2 \times 0.2/99.8$, since there are two oxygens in the molecule). The overall probability is very small since multiplying the individual probabilities together produces a number much smaller than either of them.

The relative size of the peak at $m/z$ 68 is determined using a combination of both of these probabilistic methods. Two separate entities, $^{34}S^{16}O^{18}O$ and $^{32}S^{18}O^{18}O$, contribute to $m/z$ 68, each containing two isotopes of higher atomic mass. Calculating the total probability for this combination is analogous to calculating the probability of rolling two dice so that the sum of the upper faces is 4. There are two ways to accomplish this: either a 1 and a 3 or two 2's. The probability of obtaining a 1 and a 3 in a single roll of two dice is $2 \times \frac{1}{6} \times \frac{1}{6} = \frac{2}{36} = \frac{1}{18}$, since the probability of obtaining either a 1 or a 3 is $\frac{1}{6}$, and there are two different orientations (1 and 3, or 3 and 1) that produce a sum of 4. Now, in addition to rolling a 1 and a 3, rolling two 2's also will produce a sum of 4. The probability of rolling two 2's is the same as that of rolling two 1's: $\frac{1}{6} \times \frac{1}{6} = \frac{1}{36}$. Thus, the overall probability of rolling two dice so that the sum of their upper faces is 4 is the sum of the probability of rolling a 1 and a 3 plus the probability of rolling two 2;s: $\frac{1}{18} + \frac{1}{36} = \frac{3}{36} = \frac{1}{12}$.

The relative size of the ion at $m/z$ 68, then, is determined by the sum of the probabilities for the contributing ions, which in turn are calculated from the products of the probabilities of the individual isotopes contained therein. Thus,

$$P(m/z\ 68) = P(^{34}S^{16}O^{18}O) + P(^{32}S^{18}O^{18}O)$$
$$= [P(^{34}S) \times 2P(^{18}O)] + [P(^{18}O)]^2$$

This peak is larger than that at $m/z$ 67 simply because the probability of having $^{34}S$ is greater than that of having $^{33}S$. The contributions of $^{18}O$ are so small that they essentially do not enter into consideration.

**2.2.2.2. Diazepam.** Larger molecules may contain two or more heteroatoms in addition to carbon, so that calculating probabilities for higher mass ions in the molecular ion cluster may be complicated at low mass resolution by the contributions of several different isotopes to the same ion. To illustrate, consider the molecular ion region for the tranquilizer diazepam ($C_{16}H_{13}N_2OCl$; one popular brand name is Valium). The molecular ion cluster begins at $m/z$ 283 (Figure 2.12 and Table 2.4) and is complicated even further by the fact that two separate ion clusters overlap one another—the molecular ion ($C_{16}H_{13}N_2OCl^+$) starting at $m/z$ 284 and the fragment ion resulting from the loss of hydrogen radical ($C_{16}H_{12}N_2OCl^+$) at $m/z$ 283. Sorting through this array is tedious but involves only the application of principles that were discussed earlier in this chapter.

*m/z 283.* The most intense ion in this cluster is $m/z$ 283, and its size is determined solely by the energy processes that govern fragmentation pathways, not by isotopic abundance information.

*m/z 284.* Because the ion at $m/z$ 283 contains 16 carbon and 2 nitrogen atoms, $m/z$ 284 must contain independent contributions from $^{13}C^{12}C_{15}H_{12}{}^{14}N_2{}^{16}O^{35}Cl^+$ and $^{12}C_{16}H_{12}{}^{15}N^{14}N^{16}O^{35}Cl^+$—or, for the sake of clarity, the contributions of $^{13}C$ and $^{15}N$. For 16 carbons, the *calculated* contribution is substantial ($16 \times 1.1\% = 17.6\%$), while that for 2 nitrogens is more modest ($2 \times 0.36\% = 0.7\%$). The total contributions from these ions to $m/z$ 284 is thus $17.6 + 0.7 = 18.3\%$. The *observed* intensity of $m/z$ 284 is 85.9%, indicating that 67.6% ($= 85.9 - 18.3$) of this peak comes from the actual molecular ion $^{12}C_{16}H_{13}{}^{14}N_2{}^{16}O^{35}Cl^+$.

*m/z 285.* At $m/z$ 285, things get more complicated. The $m/z$ 283 ion not only makes $^{13}C$ and $^{15}N$ contributions to the $(P + 1)^+$ ion at $m/z$ 284, but two elements from this ion make significant contributions to the $(P + 2)^+$ ion—$^{18}O$ and $^{37}Cl$. Furthermore, because of the relatively large number of carbons in the ion, the contribution of two $^{13}C$'s to the $(P + 2)^+$ ion also must be considered. All of these contributions are independent of one another (all of the contributors have different absolute masses), so that their effects are calculated separately and then added:

$$P(^{37}Cl) = 32.6\% \times 100\% \text{ (the size of } m/z \text{ 283)}/100 = 32.6\%$$

$$P(^{18}O) = 0.2\% \times 100\%/100 \qquad\qquad\qquad = 0.2\%$$

$$P(2\,^{13}C) = (16 \times 1.1)^2\%/200 \times 100\%/100 \qquad = 1.5\%$$

$$\text{Total contributions from } C_{16}H_{12}N_2OCl^+ \ = 34.3\%$$

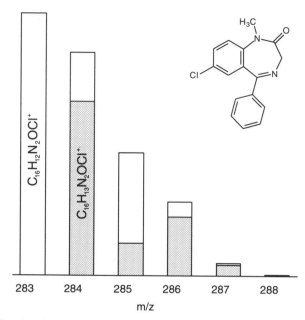

**Figure 2.12.** Overlapping ion clusters with contributions from several isotopes; the molecular ion region of diazepam.

**Table 2.4. Overlapping Ion Clusters in Diazepam (Contributions from Several Isotopes)**

| $m/z$ | $C_{16}H_{12}N_2OCl^+$ | | $C_{16}H_{13}N_2OCl^{+\cdot}$ | Intensity |
|---|---|---|---|---|
| 283 | 100.0° | | | = 100.0° |
| 284 | $^{13}C$  17.6$^c$ | + | 67.6° | = 85.9° |
|  | $^{15}N$  0.7$^c$ | | | |
| 285 | $^{37}Cl$  32.6$^c$ | + | $^{13}C$  11.9$^c$ | = 45.7$^c$ (47.1°) |
|  | $^{18}O$  0.2$^c$ | | $^{15}N$  0.5$^c$ | |
|  | $2^{13}C$  1.5$^c$ | | | |
| 286 | $^{13}C^{37}Cl$  5.7$^c$ | + | $^{37}Cl$  22.0$^c$ | = 29.0$^c$ (28.3°) |
|  | $^{13}C^{18}O$  0.03$^c$ | | $^{18}O$  0.1$^c$ | |
|  | $^{15}N^{37}Cl$  0.2$^c$ | | $2^{13}C$  1.0$^c$ | |
| 287 | $2^{13}C^{37}Cl$  0.5$^c$ | + | $^{13}C^{37}Cl$  3.9$^c$ | = 4.7$^c$ (5.0°) |
|  | $^{18}O^{37}Cl$  0.1$^c$ | | $^{15}N^{37}Cl$  0.2$^c$ (etc.) | |

°Observed intensity.
$^c$Calculated intensity.

The contributions to the $(M + 1)^+$ ion at $m/z$ 285 from $^{13}C$ and $^{15}N$ in the molecular ion are also considerable:

$$P(^{13}C) = (16 \times 1.1) \times 67.6\% \text{ (the actual size of the}$$
$$\text{molecular ion at } m/z \text{ 284)}/100 \qquad = 11.9\%$$

$$P(^{15}N) = (2 \times 0.35) \times 67.6/100 \qquad\qquad = 0.5\%$$

$$\text{Total contributions from } C_{16}H_{13}N_2OCl^+ \qquad = 12.4\%$$

Based on the observed intensities of $m/z$ 283 and 284, the *calculated* intensity of $m/z$ 285 due to the isotopic contributions from all of these ions is 34.3 + 12.4 = 45.7%. This is well within the 10% experimental error of measurement when compared to the *observed* intensity of 47.1%.

*m/z 286.* At $m/z$ 286 the contributions of coupled higher mass isotopes from the $m/z$ 283 ion begin to take effect. Since one of these elements is chlorine, the effects are not inconsequential as they were with the higher mass peaks in the sulfur dioxide molecular ion cluster. As with sulfur dioxide, the probability of each combination is calculated using the product of the individual isotopic probabilities, while the total contribution from all of these combinations to $m/z$ 286 is obtained by adding them together. Thus

$$P(^{13}C^{37}Cl) = P(^{13}C)P(^{37}Cl) = (16 \times 1.1) \times (32.6) \times 100 \text{ (the size}$$
$$\text{of } m/z \text{ 283)}/(100)^2 = 17.6 \times 32.6 \times 100/(100)^2 \quad = 5.7\%$$

$$P(^{13}C^{18}O) = P(^{13}C)P(^{18}O) - (16 \times 1.1) \times (0.2) \times 100/(100)^2 \quad < 0.1\%$$

$$P(^{15}N^{37}Cl) = P(^{15}N)P(^{37}Cl) = (2 \times 0.35) \times (32.6) \times 100/(100)^2 = 0.2\%$$

$$\text{Total contributions from } C_{16}H_{12}N_2OCl^+ \qquad\qquad = 5.9\%$$

[The $(100)^2$ factor found in the denominator of these expressions, as well as the $(100)^3$ factor in the calculations below, is necessary to adjust for the fact that percentages, rather than actual probabilities (all of which would be less than 1), are being used.]
Isotopic contributions to the $(M + 2)^+$ of the molecular ion are more important than the double isotope contributions above:

$$P(^{37}Cl) = 32.6\% \times 67.6\% \text{ (the actual size of the molecular ion)}/100 = 22.0\%$$

$$P(^{18}O) = 0.2\% \times 67.6/100 \qquad\qquad\qquad\qquad\qquad = 0.1\%$$

$$P(2^{13}C) = (16 \times 1.1)^2/200 \times 67.6/100 \qquad\qquad\qquad = 1.0\%$$

$$\text{Total contributions from } C_{16}H_{13}N_2OCl^+ \qquad\qquad = 23.1\%$$

The total *calculated* relative intensity of the $m/z$ 286 ion is the sum of all of these contributions, or 5.9 + 23.1 = 29.0%. This agrees remarkably well with the *observed* intensity of 28.3% for this ion.

*m/z 287.* By the time we get to the ion at $m/z$ 287, all contributing species involve *at least* two different isotopes. Although the list of contributors is long, only some of those ions containing $^{37}Cl$ actually make a significant contribution to the size of $m/z$ 287:

Contributors from $C_{16}H_{12}N_2OCl^+$:

$$P(2^{13}C^{37}Cl) = P(2^{13}C)\,P(^{37}Cl) = [(16 \times 1.1)^2/200] \times [32.6] \times$$
$$100/100^2 = (1.5)(32.6) \times 100/100^2 \qquad = 0.5\%$$

$$P(2^{13}C^{18}O) = P(2^{13}C)P(^{18}O) = (1.5) \times (0.2) \times 100/100^2 \qquad << 0.1\%$$
$$P(^{18}O^{37}Cl) = P(^{18}O)P(^{37}Cl) = (0.2) \times (32.6) \times 100/100^2 = 0.1\%$$

$$P(2^{15}N^{37}Cl) = P(2^{15}N)P(^{37}Cl) = (0.35)(0.35) \times$$
$$32.6 \times 100/(100)^3 \qquad << 0.1\%$$

$$P(2^{15}N^{18}O) = P(2^{15}N)P(^{18}O) = (0.35)(0.35) \times 0.2 \times$$
$$100/(100)^3 \qquad << 0.1\%$$

$$P(^{13}C^{15}N^{37}Cl) = P(^{13}C)P(^{15}N)P(^{37}Cl) = (17.6)(2 \times 0.35)$$
$$(32.6) \times 100/(100)^3 \qquad < 0.1\%$$

$$P(^{13}C^{15}N^{18}O) = P(^{13}C)P(^{15}N)P(^{18}O) = (17.6)(0.7)(0.2) \times$$
$$100/(100)^3 \qquad << 0.1\%$$

Contributors from $C_{16}H_{13}N_2OCl^{\ddagger}$:

$$P(^{13}C^{37}Cl) = P(^{13}C)P(^{37}Cl) = (16 \times 1.1) \times (32.6) \times 67.6/(100)^2$$
$$= (17.6)(32.6) \times 67.6/(100)^2 \qquad = 3.9\%$$

$$P(^{15}N^{37}Cl) = P(^{15}N)P(^{37}Cl) = (2 \times 0.35) \times (32.6) \times 67.6/(100)^2$$
$$= (0.7)(32.6) \times 67.6/(100)^2 \qquad = 0.2\%$$

$$P(^{13}C^{18}O) = P(^{13}C)P(^{18}O) = (16 \times 1.1) \times (0.2) \times 67.6/(100)^2$$
$$= (17.6)(0.2) \times 67.6/(100)^2 \qquad << 0.1\%$$

Thus the *calculated* intensity for $m/z$ 287 is 4.7%, close to the *observed* value of 5.0%.

*m/z 288 and Above.*    Although the peak at $m/z$ 288 is small, it is still not below the limits of detectability. The number of isotopic contributors from the ion at $m/z$ 283 is now substantial and includes such combinations as $^{13}C_2{}^{15}N^{18}O$, $^{13}C^{15}N_2{}^{37}Cl$, and $^{13}C^{18}O^{37}Cl$. Calculation of these intensities follows directly from previous discussions and is not pursued further.

## 2.3. EXAMPLES AND PROBLEMS—A REASONED APPROACH TO MASS SPECTRAL INTERPRETATION

This section contains three examples and seven problems designed to illustrate the material in this chapter. In six of the problems, unknown mass spectra are presented and you are asked to figure out, on the basis of the spectra alone, the structures of the organic compounds that gave rise to those spectra. Solving mass spectral unknowns is rarely trivial and is often compared to solving a puzzle. Having a rational plan for approaching unknown spectra is important—without one there is a tendency to place inappropriate weight on some of the available clues. In fact, the most common dilemma facing the mass spectral problem solver is not the lack of available clues but rather the abundance, and often seemingly contradictory nature, of those clues. The following checklist provides a plan for sorting through mass spectral information and is applicable for dealing not only with the problems in this book but also for the real-life unknowns you will encounter during the course of your work with mass spectrometry. You may want to mark this page until you are more familiar with using these steps for working problems. The first two steps apply more specifically to real-life unknowns, although in a few cases some chemical information will be provided (and should not be ignored!) with the problems in this book.

1. Do a library search on the spectrum, if at all possible. Use all of the mass spectral library resources at your disposal. *Obtain actual printouts* of the spectra picked by the library and *compare them carefully* with the unknown spectrum. The importance of doing an actual visual comparison of the unknown and standard spectra cannot be overemphasized. If the match is not extremely good, assume the spectrum picked by the library may not be the correct one.

2. Get as much chemical information about the compound as you can. Where did it come from? What types of compounds are likely to come from such a source? The more information you can get, the narrower your list of possible structures will be.

3. Try to identify the molecular ion, or decide whether it is present at all in the spectrum. This is often the most critical step in solving an unknown—the structure cannot be determined if you do not know the molecular weight! Unfortunately, it is also one of the most difficult steps because some compounds do not give molecular ions in electron ionization mass spectrometry. To verify the reasonableness of your choice:

   a. Check isotopic abundances. If the $(M + 1)^+$ ion is too large to be accommodated by a reasonable number of carbons, the "$(M + 1)^+$" ion itself may be the molecular ion (Section 2.2.1.3).

   b. Determine the first losses from the proposed molecular ion. Some losses are virtually impossible, such as those of 12, 14, or 23 daltons (see Chapter 4). The presence of such ions in the spectrum will make your choice more difficult to support.

   c. Does the spectrum appear to be "dirty" (i.e., are there lots of small, extraneous peaks, even at very high masses)? If so, the possibility of missing a weak molecular ion hidden in the "grass" becomes a real problem.

   d. If GC data for the compound is available (e.g., by GC/MS), compare the proposed molecular weight of this compound with that of compounds that elute at similar retention times. For the most part, you would not expect a compound having a molecular weight of 175 to elute at a retention time near that of one with a molecular weight of 300.

4. Is the molecular weight even or odd? Because of the combined odd valence and even atomic weight of nitrogen, the empirical formula of any organic compound containing only C, H, N, O, Si, S, P, or halogen that has an odd

number of nitrogen atoms will have an odd molecular weight. Similarly, an empirical formula having an even number of (or no) nitrogen atoms will have an even molecular weight. This rule, called the *odd-nitrogen rule* (Section 4.3), is applicable only to odd-electron ions (Section 3.1) and can be used with certainty only with the molecular ion.

5. Examine the molecular ion cluster and other major ion clusters in the spectrum for isotopic abundance information. Look for heteroatom patterns, especially those of the halogens and sulfur. Try to calculate the number of carbons, oxygens, and so forth, remembering the limitations of such calculations (Section 2.2.1.3).

6. What does the overall appearance of the spectrum tell you? Is one bond particularly fragile? Is the compound likely to be aromatic or aliphatic? These issues will be discussed in Chapter 4.

7. Look for low mass ion series (there may be more than one—see Chapters 4 and 5). Do the library search results give any clues as to the family of compounds with which you may be dealing?

8. Make a list of suggested losses from the molecular ion and try to make a pattern of them. Have you encountered a similar pattern of losses from other compounds?

9. Look for intense odd-electron ions in the spectrum (see Section 3.1, but note this is next to impossible with compounds containing nitrogen!). These often provide clues to structural arrangements that fragment in a specific manner (see, e.g., the γ-hydrogen rearrangement and retro Diels–Alder fragmentation in Chapter 6).

10. Condense all of this information and speculate on a structure. To aid in this process, it is helpful to remember that the total number of rings plus double bonds in a molecule containing $x$ carbon atoms, a total of $y$ hydrogen and/or halogen atoms, $z$ nitrogens and $n$ atoms of oxygen and/or sulfur (general formula $C_xX_yN_zO_n$) is given by the formula

$$\text{Total rings plus double bonds} = x - \tfrac{1}{2}y + \tfrac{1}{2}z + 1 \tag{2.7}$$

This formula works for the molecular ion and, indeed, all odd-electron ions (see Section 3.1). Once you have arrived at a structure, are all of the major ions explainable on the basis of this structure? Can you write reasonable mechanisms for fragmentations leading to particularly stable ion products (see Chapter 7)? If not, be skeptical. Try different arrangements of the same functional groups. Sometimes an isomer of the original proposal contains a key element that causes the whole spectrum to make sense.

11. Above all, do not give up. Solving mass spectral unknowns with limited additional information is often impossible, even for "old hands." Keep trying.

**Examples 2.1 to 2.3.** Identify the compounds whose spectra are shown in Figures 2.13, 2.14, and 2.15.

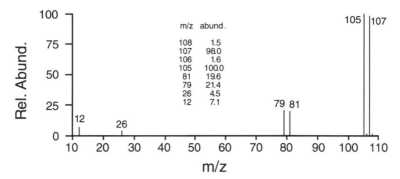

**Figure 2.13.** Mass spectrum for Example 2.1.

*Answer 2.1.* There are no library search results or chemical histories given for this unknown, and there is not enough information to evaluate the choice for molecular ion. The feature of this spectrum that immediately grabs our attention is the pair of doublets at high mass, with the peaks in each separated by 2 daltons. This pattern is consistent *only* with the presence of one bromine in the molecule, an observation confirmed by the fact that the lower doublet occurs at $m/z$ 79 and 81 ($Br^+$).

The molecular ion occurs at $m/z$ 105 (not 107!), which is an odd mass. Because of the "odd nitrogen rule" (guideline 4 above), this means that the compound *must* contain an odd number of nitrogens; in this case more than one is impossible because of the low molecular weight. The combination of one bromine and one nitrogen accounts for 93 of the 105 daltons in the molecular weight, leaving only 12 daltons unassigned. The only realistic choice, one carbon, is confirmed by the sizes of the ions at $m/z$ 106 and 108. Each of these ions has an intensity of approximately 1.5% relative to the ion just below it in mass, consistent with the presence of exactly one carbon (1.1%) and one nitrogen (0.4%). The "ion" at $m/z$ 106 is thus due to the combined presence of $^{13}C^{14}N^{79}Br$ and $^{12}C^{15}N^{79}Br$, while that at $m/z$ 108 is due to $^{13}C^{14}N^{81}Br$ and $^{12}C^{15}N^{81}Br$.

A likely structure for this compound is $N{\equiv}C{-}Br$, cyanogen bromide. Loss of 26 daltons from the molecular ion as $\cdot CN$ produces the $Br^+$ ion at $m/z$ 79 and 81 [Eq. (2.8); the single-headed or fishhook arrow convention is explained fully in Section 3.1]. The only other ions in the spectrum occur at $m/z$ 26 ($NC^+$) and 12 ($C^+$). The structure and spectrum thus appear to be consistent with one another (cyanogen bromide; NCBr):

$$Br{-}C{\equiv}\overset{+}{N} \xrightarrow{\text{- Br}^\bullet} \oplus C{\equiv}N \longrightarrow \overset{+}{\cdot}C \quad ; \quad \overset{+}{Br}{-}C{\equiv}N \xrightarrow{\text{- CN}^\bullet} Br{\oplus} \quad (2.8)$$

$m/z$ 105 $\qquad$ $m/z$ 26 $\qquad$ $m/z$ 12 $\qquad\qquad$ $m/z$ 105 $\qquad$ $m/z$ 79

**2.2.** There are several small ions in this spectrum between $m/z$ 98 and 102, and it is not clear which of them might be the molecular ion. Since these ions are separated by 2 daltons, this pattern looks suspiciously like an isotope cluster produced by more than one chlorine (Figure 2.6). In fact the relative ratio of these ions (3.5:2.4:0.4 =

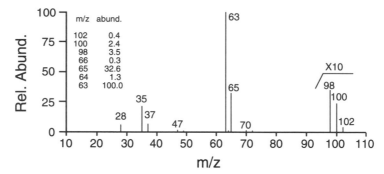

**Figure 2.14.** Mass spectrum for Example 2.2. In the graphical representation, the peaks at $m/z$ 98 to 102 are shown 10 times their actual size.

100:69:11) is close to that expected for two chlorines (Section 2.2.1.1 and Figure 2.5c). Confirmation of this is found in the base peak at $m/z$ 63, which is 35 daltons (the mass of one chlorine) less than $m/z$ 98 and which also has an associated peak at $m/z$ 65 that is about 33% of its intensity.

This spectrum is similar to that of 1,2-dichloroethylene (Figure 2.5c) in several respects. Both compounds have base peaks containing one chlorine, small fragment ions in the area of $m/z$ 47 to 50 and $m/z$ 35 to 37, and similar intensity patterns in the molecular ion cluster. Since the molecular weight of this compound is 2 daltons higher than that of dichloroethylene, an isomer of dichloroethane ($C_2H_4Cl_2$) is a reasonable possibility.

However, the size of the $m/z$ 64 ion (1.3%) is only large enough to accommodate the presence of one carbon in $m/z$ 63, and since $m/z$ 63 appears to result from loss of chlorine radical from $m/z$ 98, this means the molecular ion contains only one carbon as well. (Note that the $^{13}C$ isotope peaks in the molecular ion region are too small to be observed, so that we must *infer* the number of carbons in this ion based on the composition of its fragment ions and proposed neutral losses.) A group that satisfies this criterion and also has the same mass as the $-CH_2CH_2-$ group of dichloroethane is a carbonyl group. Another possiblity, $Cl-N=N-Cl$, can be ruled out by the relative abundance of the $m/z$ 64 ion, which is too large for even two nitrogens. All aspects of this spectrum, including the small fragment ion at $m/z$ 28 ($CO^+$), fit a compound having two chlorines attached to a central CO:

$$Cl_2^{+\cdot} \xleftarrow{-CO} \overset{O}{\underset{Cl}{\underset{}{\overset{}{\overset{\parallel}{\underset{Cl}{}}}}}}^{+\cdot} \xrightarrow[a]{-Cl^\cdot} Cl-C\equiv O^{\oplus} \qquad (2.9)$$

$m/z$ 70      $m/z$ 98      $m/z$ 63

$\overset{\oplus}{C}Cl \qquad\qquad Cl^{\oplus} \qquad\qquad CO^{+\cdot}$

$m/z$ 47      $m/z$ 35      $m/z$ 28

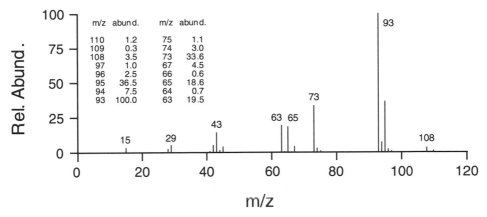

| m/z | abund. | m/z | abund. |
|-----|--------|-----|--------|
| 110 | 1.2 | 75 | 1.1 |
| 109 | 0.3 | 74 | 3.0 |
| 108 | 3.5 | 73 | 33.6 |
| 97 | 1.0 | 67 | 4.5 |
| 96 | 2.5 | 66 | 0.6 |
| 95 | 36.5 | 65 | 18.6 |
| 94 | 7.5 | 64 | 0.7 |
| 93 | 100.0 | 63 | 19.5 |

**Figure 2.15.** Mass spectrum for Example 2.3.

In confirmation of this, this spectrum lacks ions showing the presence or loss of hydrogen—ions that are present in the dichloroethylene spectrum (Figure 2.5c). In particular, ions corresponding to those at $m/z$ 59 and 60 due to hydrogen loss from $m/z$ 61, $m/z$ 48 from $CH^{35}Cl^{+}$, and $m/z$ 36 due to $H^{35}Cl^{+}$ are missing. This is inconsistent with dichloroethane but does support the presence of a carbonyl group (phosgene; $COCl_2$).

*2.3.* This is a difficult spectrum for the novice to interpret. However, by careful application of the guidelines at the beginning of this section, the answer becomes apparent after a little work.

The molecular weight of this compound appears to be even (108), so that nitrogen is not suspected (odd-nitrogen rule). The isotope clusters at $m/z$ 108 to 110 and $m/z$ 93 to 97 suggest the presence of one chlorine (especially the pair of ions at $m/z$ 93 and 95). The ions between $m/z$ 63 and 67 are confusing—the pattern looks somewhat consistent with that for three chlorines, but this is impossible given the masses of the ions. The isotopic abundance information in $m/z$ 94 is also confusing, indicating the presence of approximately seven carbons in addition to the chlorine. This also is impossible since an ion at $m/z$ 93 could contain no more than one chlorine and five carbons.

The isotope cluster at $m/z$ 73 to 75, which seems to be due to a fragment ion that has lost the chlorine ($108 - 35 = 73$), is more informative. Once again the $(P + 1)^{+}$ ion appears to contain more carbons than is possible for the mass of the ion ($3.0/33.6 = 8.9\% \Rightarrow 8$ carbons), *but* the $(P + 2)^{+}$ ion ($1.1/33.6 = 3.3\%$) is also much too large to be due to contributions from $^{13}C$ alone. In fact, it is too large to accommodate either $^{13}C$ or $^{18}O$ and too small for either $^{37}Cl$ or even $^{34}S$! What elements are left? From Table 2.1, we see that this isotope pattern fits that of silicon, an element mentioned only briefly up to this point.

If $m/z$ 73 contains one silicon, we expect $^{29}Si$ to contribute $4.7/92.2 = 5.1\%$ to $m/z$ 74 and $^{30}Si$ to contribute $3.1/92.2 = 3.5\%$ to $m/z$ 75. One silicon alone thus accounts for the abundance of the $m/z$ 75 ion, but leaves $8.9 - 5.1 = 3.8\%$ intensity unaccounted for in $m/z$ 74—three carbons could make up this difference. Three carbons and a silicon have a combined mass of $(3 \times 12) + 28 = 64$ daltons, leaving nine hydrogens to make up the missing mass, and we arrive at $^+Si(CH_3)_3$ as a likely structure for $m/z$ 73 (other isomeric structures are possible, but see Section 2.2.1.6).

The relative intensities of the ions at $m/z$ 93 to 97 are now understandable, if we assume that the molecular ion $[Cl-Si(CH_3)_3{}^+]$ loses methyl radical (15 daltons) to form $m/z$ 93:

$m/z\ 94 \Rightarrow 5.1\%\ (^{29}Si) + 2.2\%\ (^{13}C)$          $= 7.3\%\ (7.5\%\ \text{obs.})$

$m/z\ 95 \Rightarrow 32.0\%\ (^{37}Cl) + 3.5\%\ (^{30}Si)$         $= 35.5\%\ (36.5\%\ \text{obs.})$

$m/z\ 96 \Rightarrow [32.0 \times 5.1\ (^{37}Cl^{29}Si)]\% + [32.0 \times$
     $2.2\ (^{37}Cl^{13}C)]\% = 1.6 + 0.7$          $= 2.3\%\ (2.5\%\ \text{obs.})$

$m/z\ 97 \Rightarrow [32.0 \times 3.5\ (^{37}Cl^{30}Si)]\%$          $= 1.1\%\ (1.0\%\ \text{obs.})$

Other isotopic combinations make only small contributions to the sizes of these peaks.

The enigmatic group of ions at $m/z$ 63 to 67 must be due to ions retaining chlorine. If this is true for $m/z$ 63, then $m/z$ 65 must have a corresponding $19.6 \times 32.0 = 6.2\%$ contribution from $^{37}Cl$. Ignoring any possible contributions from silicon at this point, this leaves $18.6 - 6.2 = 12.4\%$ remaining for the actual size of $m/z$ 65, which in turn must produce a $12.4 \times 32.0 = 4.0\%$ ion at $m/z$ 67. This is close to what is observed. The abundances of the small ions at $m/z$ 64 and 66 in this cluster are too small to be known with certainty, so whether the remaining mass in these ions is due to silicon or carbon remains unresolved. Because of the lower electronegativity of silicon, we might suspect that it stabilizes the positive charge better than carbon (Chapter 5). Equation (2.10) offers plausible structures for the major ions in the spectrum [trimethylchlorosilane; $ClSi(CH_3)_3$]:

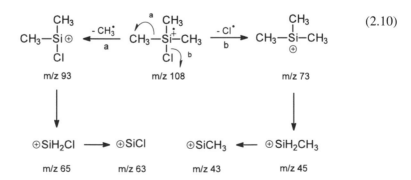

(2.10)

**Problem 2.5.** The molecular ion region in the mass spectrum of an unknown organic compound shows the following isotope cluster:

$m/z\ 123 - 42.3\%$ (relative to a base peak at $m/z$ 77)
$124 - 3.0\%$
$125 - 0.3\%$

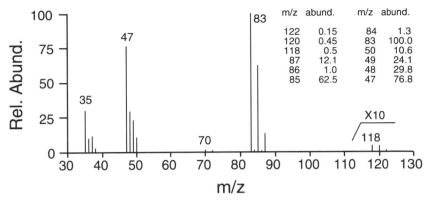

**Figure 2.16.** Mass spectrum for Problem 2.6.

Assuming that this data is well within experimental error, how much can you deduce about the structure of this compound? (Knowledge of the base peak is helpful but not necessary for solving the problem. Focus first on the isotopic abundance data.)

***Problems 2.6 through 2.10.*** The spectra for these unknowns are given in Figures 2.16 to 2.20. The ions at $m/z$ 118 to 122 in Figure 2.16 are shown 10 times larger than their actual intensities. The observed abundances for these ions are given in the tabular data in the figure.

***Problem 2.11.*** The mass spectrum shown in Figure 2.21 was obtained from the "cut," or diluting agent, of an illicit methamphetamine sample (the structure of methamphetamine is given in Section 5.3). What is the structure of this unknown compound?

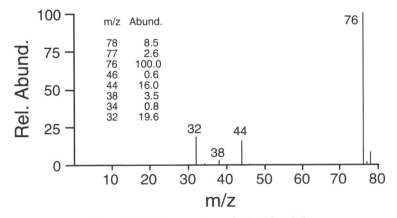

**Figure 2.17.** Mass spectrum for Problem 2.7.

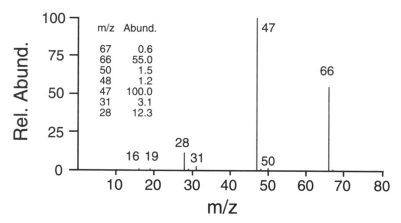

**Figure 2.18.** Mass spectrum for Problem 2.8.

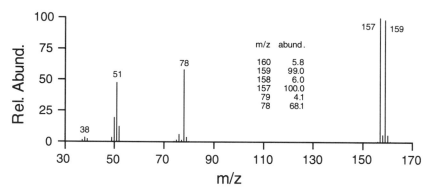

**Figure 2.19.** Mass spectrum for Problem 2.9.

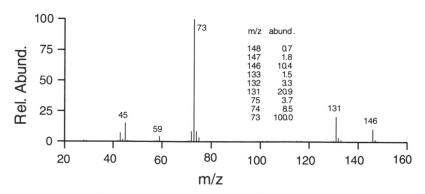

**Figure 2.20.** Mass spectrum for Problem 2.10.

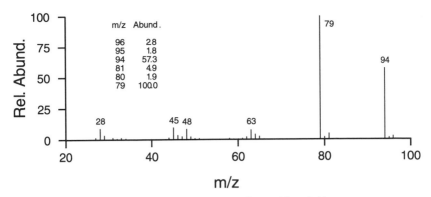

**Figure 2.21.** Mass spectrum for Problem 2.11.

# CHAPTER 3

# IONIZATION, FRAGMENTATION, AND ELECTRON ACCOUNTING

## 3.1. EVEN- AND ODD-ELECTRON SPECIES

When an electrically neutral molecule is bombarded with electrons in the ion source, one of the molecule's electrons is stripped away, leaving an unpaired electron and a positive charge where the lost electron used to be. The radical ion concept may be unfamiliar at first, but as one begins to write detailed mechanisms for mass spectral fragmentations, it becomes readily apparent that keeping track of each electron is important.

To make this task easier, let us begin by considering which ions and neutral species have even or odd numbers of valence electrons. To fully grasp this point, beginners may want to write out complete valence electronic structures of molecules. It is important to remember that both ions and neutral fragments can be either odd- or even-electron. A few simple examples should suffice before extrapolating to more complex structures (Figure 3.1).

*Even-electron neutral fragment.* Nearly all electrically neutral organic molecules have an even number of valence electrons in their ground state. Most even-electron species have all of their electrons paired either in single, double, or triple bonds or as nonbonded (lone) pairs. A few even-electron neutral fragments also exist as diradicals, that is, with two unpaired electrons located on different atoms; but these occur infrequently because they are inherently less stable than molecules in which all of the electrons are paired. The valence electronic structure for the ground state of formamide ($HCONH_2$), a molecule containing a carbon–oxygen double bond and two heteroatoms with nonbonded lone pairs of electrons, is shown at the top of Figure 3.1. Even-electron neutral molecules in their ground state will often be denoted by the superscript ° following their formulas.

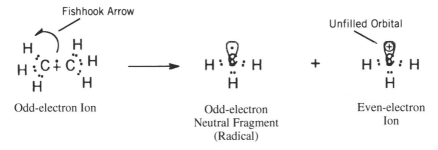

Figure 3.1. Electronic structures of some simple molecules, radicals, and ions.

*Odd-electron ion (radical ion).* Ionization of an even-electron neutral molecule by ejection of one electron forms an odd-electron ion since the single electron remaining in the now partially filled orbital is unpaired. Because we will only deal with even-electron neutral molecules in this book, the molecular ions of all of these compounds will be odd-electron ions. The molecular ion for ethane ($C_2H_6$) is seen at the left of the equation at the bottom of Figure 3.1. Odd-electron ions are denoted by the symbols $\bar{+}$ or $^+$·, placed either after the formula or at the actual site of ionization. If the radical site and the site of positive charge are located on separate atoms, the · may be placed next to the atom having the radical site, and either a + or a ⊕ placed next to the atom having the positive charge.

*Odd-electron neutral fragment (radical).* When an odd-electron ion fragments, it must necessarily produce at least one odd-electron fragment in order to balance the total number of electrons. If exactly two products are formed (which is the case for most fragmentations), then one of the products will be an odd-electron species and the other an even-electron species. In Figure 3.1, the central bond of the molecular ion of ethane, already weakened by the loss of a bonding electron, breaks apart. One electron, denoted in Figure 3.1 by a single-headed or *"fishhook" arrow,* stays with one of the methyl groups (arbitrarily assigned to the left one in Figure 3.1) to produce a methyl radical ($CH_3$·). Methyl radical contains one unpaired electron, located in a nonbonding $sp^3$ orbital on the carbon.

Methyl radical is also electrically neutral. The valence shell for atomic carbon contains four electrons, so that carbon will remain electrically neutral if it has four electrons associated with it. In methyl radical the carbon shares an electron pair with each of the three hydrogens and in the process "owns" one electron from each bonding pair, for a total of three bonding electrons. The fourth electron is the nonbonded unpaired electron, which completes the requirements for carbon's electrical neutrality. Each hydrogen is electrically neutral because it also owns one electron (one-half of an electron pair shared with carbon), thus fulfilling its normal valence requirement. Because methyl radical is neutral, it is not detected by the mass spectrometer.

*Even-electron ion.* The other product resulting from fragmentation of the ethane molecular ion is methyl carbocation ($CH_3^+$), shown at the right of the equation at the bottom of Figure 3.1. The main difference between methyl carbocation and methyl radical is the absence of the unpaired electron in the nonbonding orbital associated with carbon. Since the ion lacks this electron, it falls one electron short of producing electrical neutrality on carbon and is thus positively charged. At the same time, however, all of the valence electrons in methyl carbocation are paired in the bonds between carbon and the hydrogens, leaving the nonbonding orbital completely empty and rendering the ion an even-electron species.

The fragmentation depicted at the bottom of Figure 3.1 shows an odd-electron ion dissociating to give an odd-electron neutral fragment (radical) and an even-electron ion, a process we will symbolize by the equation

$$OE^+ \rightarrow OE\cdot + EE^+$$

where OE and EE stand for odd- and even-electron species, respectively.

Odd-electron ions also can fragment to give a second odd-electron ion, but in that case the other product must be a neutral molecule since the unpaired (odd) electron is now found in the ionic fragment:

$$OE^+ \rightarrow OE^+ + EE^\circ$$

where the $^\circ$ superscript denotes electral neutrality. Whereas formation of a radical and an even-electron ion from an odd-electron ion often occurs with cleavage of a single bond (as illustrated in Figure 3.1), formation of a second odd-electron ion and a neutral molecule is a more complex process and must involve both bond breaking and new bond formation—in other words, rearrangement of the atoms of the original ion. This type of fragmentation may be sensitive to specific structural arrangements in the molecule and, as a result, can be very useful for detecting these arrangements in molecules.

Even-electron ions, formed by fragmentation of either odd- or other even-electron ions, also have two ways of distributing valence electrons during their own dissociation. In either case, the two products must both be of the same type—either both even- or both odd-electron:

$$EE^+ \rightarrow OE^+ + OE\cdot$$

or

$$EE^+ \rightarrow EE^+ + EE^\circ$$

Both processes are commonly observed. The first may occur with or without *electronic* rearrangment; that is, it may proceed either by (a) breaking only one single bond in the ion or (b) first breaking a single bond, then reorganizing the remaining electronic structure of the molecule, especially if the molecule contains a number of conjugated double bonds. The second process, as with the formation of neutral molecules from odd-electron ions, usually proceeds with some structural rearrangement, although these rearrangements often are less useful for determining molecular structure. The formation of doubly and triply bonded small molecules (such as $CH_2{=}CH_2$, $CH_2{=}O$, $HC{\equiv}CH$, $HC{\equiv}N$, $C{\equiv}O$, or $N_2$) is often a strong driving force for this type of fragmentation. We will see many examples in subsequent chapters.

Each of the fragmentation processes listed above can occur, at least theoretically, either by having the positive charge remain in the same part of molecule in which it originally resided (*charge retention*) or by moving the charge to a new site in the molecule with loss of the original ionization site (*charge migration*). Depending on the ion product, even- and odd-electron ions accomplish this in different ways, some involving the movement of a single electron, others with the movement of an electron pair. We need different notations to distinguish between these two processes, thus we denote movement of individual electrons with the use of single-headed or "fishhook" arrows ($\rightharpoonup$) and movement of electron pairs with full-headed arrows ($\rightarrow$).

## 3.2. SITE OF INITIAL IONIZATION

Ionization in mass spectrometry occurs by a complex process in which some or all of the energy of the bombarding electron is transferred to the sample molecule—enough energy to overcome the ionization potential of the molecule and to cause the molecule to eject an electron and form a positive ion. In fact, the energy of the bombarding electrons (70 eV) is more than three to four times the energy needed to break even the strongest bonds in most organic molecules so that, if a substantial proportion of this energy is transferred to the molecule, additional ionization (formation of doubly and triply charged ions) and/or fragmentation can occur with little provocation. The electrons most vulnerable to ejection are those in molecular orbitals having the highest energy (the *highest occupied molecular orbitals,* or HOMOs). Indeed, the *ionization potential* of the molecule is defined as the energy needed to completely remove one of these electrons from the molecule. These orbitals also give rise to the weakest bonds in the molecule. Figure 3.2 shows the relative energies of different types of molecular orbitals found in organic compounds. Basically, these molecular orbitals fall into five categories: $\sigma$, $\pi$, $n$, $\pi^*$, and $\sigma^*$.

The strongest bonds in the molecule are the $\sigma$-bonds—those in which two adjacent atoms share a single pair of valence electrons in a "head-on" overlap of electron density. Molecular orbitals describing $\sigma$-bonds in the molecule are called $\sigma$-orbitals. There are as many $\sigma$-orbitals as there are $\sigma$-bonds in the molecule; each orbital is filled with a pair of electrons. It is important to remember that, although the electron distribution described by each molecular orbital may be highly concentrated between two atoms (consistent with what we picture as a "single bond"), the actual probabil-

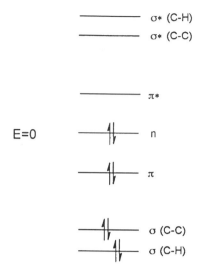

**Figure 3.2.** Relative energies of molecular orbitals in organic compounds.

ity of finding an electron in that orbital is spread out unevenly over the entire molecule.

Bonds in which electron density overlap is achieved in a "side-by-side" manner (as with the $p$-orbitals of the carbons in a benzene ring) are called $\pi$-bonds, and molecular orbitals describing the $\pi$-bonds in a molecule are denoted as $\pi$-orbitals. Because electron density overlap in $\pi$-bonds is less direct than it is in $\sigma$-bonds, $\pi$-bonds are weaker than $\sigma$-bonds. This means that $\pi$-orbitals are higher in energy than $\sigma$-orbitals. (Do not confuse this with the fact that double bonds are stronger than single bonds! A double bond contains *both* a $\pi$-bond and a $\sigma$-bond.) As with $\sigma$-orbitals, there are as many $\pi$-orbitals as there are $\pi$-bonds in the molecule, and each orbital contains a pair of electrons. $\pi$-Orbitals in some molecules (aromatic compounds in particular) are more apt than $\sigma$-orbitals to have the electron density spread out over several atoms in the molecule.

Most heteroatoms in neutral molecules, unlike carbon and hydrogen, do not share all of their valence electrons with other atoms in chemical bonds. Instead they have one or more "*nonbonding*" orbitals, each containing a pair of electrons called a "lone pair." In molecules containing heteroatoms, these molecular orbitals are denoted as $n$-orbitals, and, because they are not involved in bonds with other atoms, no bonding energy is gained. These orbitals thus are found near the "zero" of energy. In contrast to the $\sigma$- and $\pi$-orbitals, which may describe bonding that extends over several atoms at the same time, $n$-orbitals remain essentially localized on the individual heteroatoms.

Located at even higher energies are "antibonding" $\sigma^*$- and $\pi^*$-orbitals. There are as many $\sigma^*$- and $\pi^*$-orbitals as there are $\sigma$- and $\pi$-orbitals, and, generally speaking, they are as far above the zero of energy as the $\sigma$- and $\pi$-orbitals are below it. Antibonding orbitals are empty in the ground state, although irradiating the molecule with ultraviolet or visible radiation can promote a bonding or nonbonding electron into an antibonding orbital, causing the molecule to be in an excited state. Excited-state chemistry falls under the realm of organic photochemistry and generally has little application in electron ionization mass spectrometry.

Because ionization in mass spectrometry usually causes ejection of an electron from one of the highest energy filled orbitals in the molecule, the order of orbital energy shown in Figure 3.2 makes clear that, if the molecule contains heteroatoms, ionization should occur preferentially at one of the $n$-orbitals on the heteroatoms, and the resulting positive charge can be considered to be localized exclusively on that heteroatom, at least initially. For our purposes, all heteroatoms can be treated equally in this regard; a molecule containing several different heteroatoms may show evidence of ionization at each of them.

If a molecule contains no heteroatoms, but does have carbon–carbon double or triple bonds, the ejected electron comes from the highest energy $\pi$-orbital(s). Since these orbitals may show bonding between several contiguous atoms at the same time (Figure 3.3), the positive charge may not be localized on a single atom (especially in

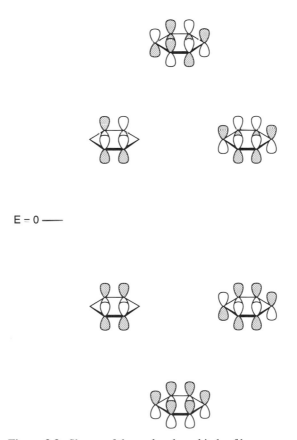

**Figure 3.3.** Shapes of the molecular orbitals of benzene.

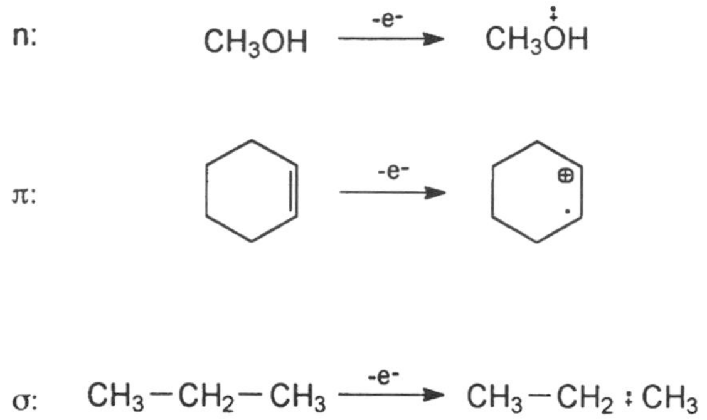

n:   $CH_3OH \xrightarrow{\ \ -e^-\ \ } CH_3\overset{\cdot\cdot}{O}H$

π:

σ:   $CH_3-CH_2-CH_3 \xrightarrow{\ \ -e^-\ \ } CH_3-CH_2 \overset{\cdot}{\underset{\cdot}{}} CH_3$

**Figure 3.4.** Examples showing notation for localization of initial ionization site.

aromatic rings). Saturated hydrocarbons, lacking both n- and π-orbitals, must lose an electron from one of the σ-orbitals. We will use the symbolism shown in Figure 3.4 to designate these various types of ionization.

## 3.3. ENERGY CONSIDERATIONS IN FRAGMENTATION PROCESSES

The most difficult task in mass spectral interpretation is trying to *predict* how a given molecule will fragment under electron ionization conditions. In fact, our understanding of the myriad processes that can take place under these conditions is so limited that rationalization of the fragmentation pattern after having observed it, rather than predicting it beforehand, is often the only realistic approach. On the other hand, if there were not some unifying concepts to the fragmentation of organic molecules, this book could contain only a list of molecules and their fragmentation processes.

The course of any chemical reaction is governed either by thermodynamic factors, in which the relative stabilities of the products and reactants determine the amounts of each present in the final mixture, or by kinetic factors, in which the relative rates of various possible reactions of the starting material determine product distribution. To be governed by thermodynamic control, the reactants and products must reach a state of equilibrium in which all products can revert to starting materials by reversing the reaction pathways that led to their formation. The strong electric fields present in the ion source make equilibrium between fragmenting ions and their products virtually inconceivable; one of the main functions of the ion source is to *remove* the ion products as rapidly as they are formed! It seems most likely, then, that mass spectral fragmentations are controlled by the relative rates at which various fragmentations occur.

The rate of a chemical reaction is given by the Arrhenius equation

$$k = Ae^{-\Delta G\ddagger/RT} \tag{3.1}$$

where $k$ is the rate constant for the reaction, $A$ is a "frequency factor" determined by the nature of the reaction, $\Delta G^{\ddagger}$ is the free energy of activation, $R$ is the gas constant, and $T$ is the temperature at which the reaction takes place. The larger the value of $k$, the faster the reaction. At constant $T$, as $\Delta G^{\ddagger}$ gets smaller, $e^{-\Delta G^{\ddagger}/RT}$ approaches $e^{0} = 1$ and $k \rightarrow A$—that is, the reaction occurs more easily. On the other hand, as $\Delta G^{\ddagger}$ gets larger, $e^{-\Delta G^{\ddagger}/RT}$ approaches $e^{-\infty} = 0$ and $k \rightarrow 0$, and the reaction occurs with increasing difficulty.

Provided the product ions do not fragment as soon as they are formed, those reactions in the ion source having the largest values of $k$ generate the most product ions per unit time. As a result, greater numbers of those quickly formed ions reach the electron multiplier, and the peaks in the spectrum corresponding to them are more intense. The actual timing of various fragmentation processes is critical since ions leave the ion source within approximately $10^{-5}$ s of being formed. Ions that are relatively stable (that is, have large $\Delta G^{\ddagger}$'s for further fragmentation) will react only slowly and tend to remain intact until they reach the detector. Less stable ions dissociate to varying degrees before they leave the ion source, decreasing the number of these ions reaching the electron multiplier.

Determining the relative rates of various fragmentations hinges mostly on estimations of $\Delta G^{\ddagger}$, the energy needed to boost an ion to the *transition state* for the reaction, a configuration of maximum energy in which breaking bonds are severely stretched and any new bonds are just starting to form (Figure 3.5). This state is differentiated from a *reaction intermediate,* which occurs at an energy minimum (even

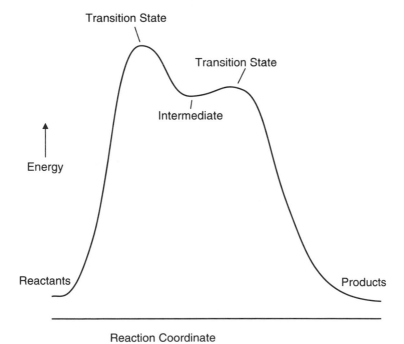

**Figure 3.5.** Energy diagram for a hypothetical reaction having an intermediate and two transition states.

if that minimum is substantially above the energies of the reactants and products). Because energy is required to change the configuration of the intermediate, an intermediate has a finite lifetime whereas a transition state does not.

The equation for determining $\Delta G^{\ddagger}$ is

$$\Delta G^{\ddagger} = \Delta H^{\ddagger} - T \Delta S^{\ddagger} \tag{3.2}$$

where $\Delta H^{\ddagger}$ is the change in enthalpy or heat of activation (the energy needed to stretch and twist bonds toward their breaking point) and $\Delta S^{\ddagger}$ is the change in entropy (or orderliness) of the entire system (the more order required, the more *negative* the value of $\Delta S^{\ddagger}$). The term $\Delta H^{\ddagger}$ reflects changes in the heats of formation required to make the reactions occur, and is affected by bond strengths and the relative stabilities of reactants and products. Although breaking bonds in the reactants obviously requires an input of energy (raising $\Delta H^{\ddagger}$), the simultaneous formation of bonds in the products (as happens during rearrangement reactions) can lower $\Delta H^{\ddagger}$ sufficiently to allow a reaction to occur that would otherwise be entropically unfavorable (see below).

On the other hand, $\Delta S^{\ddagger}$ is a measure of how difficult it is to get atoms to align themselves so that the reaction can take place. A reaction that requires a molecule simply to fall apart (e.g., stretch a bond until it breaks) may have little effect on $\Delta S^{\ddagger}$ since the alignment of atoms in the transition state is, if anything, slightly more random than in the reactant. On the other hand, the need to arrange several atoms in a geometrical pattern, as occurs during rearrangment reactions, leads to a large negative value for $\Delta S^{\ddagger}$ and increases the overall energy of activation. All other things being equal, the change in entropy required to form three-, four- and six-membered rings during rearrangement reactions is less than for formation of five-, seven-, and larger-membered rings. Many of the important rearrangements discussed in later chapters proceed via cyclic transition states involving three-, four-, and six-membered rings.

The energy diagram in Figure 3.6 describes the hypothetical fragmentation of a molecule that yields three *observed* ions. In this illustration, $M^{+}$ is shown to be less stable than the final ion product $F^{3\oplus}$. Although this is often the case, molecular ions that are unusually stable occur frequently with aromatic compounds and may, in fact, be lower in internal energy than all of their fragment ions. For all of the fragmentations of these compounds $\Delta G^{\ddagger}$ may be so large that fragmentation occurs only infrequently. This will be reflected in a mass spectrum in which most of the observed ions are unfragmented molecular ions.

Because electron ionization mass spectrometry provides so much energy for ionization, it is possible to overcome even highly unfavorable energy barriers and form intermediates and final ion products that are relatively unstable. Processes requiring high-energy input occur more infrequently than those requiring less energy, and only small numbers of ions corresponding to these fragmentations will be seen. Nonetheless, for any given molecule, the relative intensities of fragment ions observed in the spectrum are reflections of the *relative* energies needed to cause various fragmentations to occur. It is this fact that makes predicting fragmentation processes in mass spectrometry so difficult.

E=70eV  - - - -

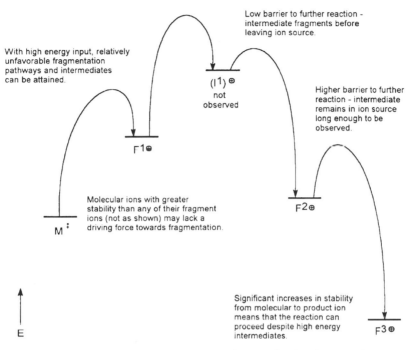

**Figure 3.6.** Some energy considerations affecting how molecules fragment in mass spectrometry.

The energy barriers for some fragmentations, such as that leading from $I^{1\oplus}$ to $F^{2\oplus}$ in Figure 3.6, may be so low that the lifetime of the intermediate ion is too short to allow it to exit the ion source before fragmenting further. These ions will not be observed despite the fact that they are formed initially. Section 8.4 provides evidence that the hallucinogenic drug phencyclidine fragments via an expected intermediate ion that is not observed in the spectrum.

Consideration of the α-cleavage of the stimulant drug amphetamine, which is covered in detail in Section 5.3, should help clarify these ideas. The energy diagram for this fragmentation is shown in Figure 3.7. The fragmentation itself, which is typical of all aliphatic amines, involves initial ionization at the nitrogen atom (Section 3.2) followed by cleavage of bonds to the carbon next to the nitrogen, but not including the bond between that carbon and the nitrogen (see Chapter 5). This fragmentation can occur either without moving the positive charge from the nitrogen (charge retention) or by movement of the charge to another part of the molecule (charge migration).

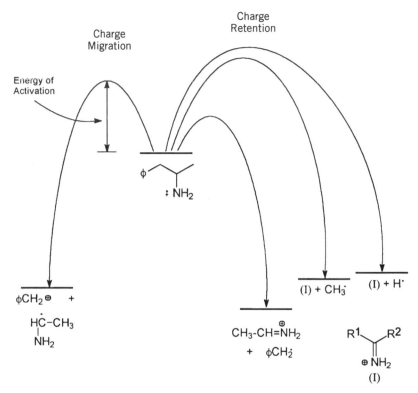

**Figure 3.7.** Energy diagram showing relative transition-state energies for various α-cleavage fragmentations of amphetamine.

There are six different pathways by which amphetamine can undergo α-cleavage (Table 3.1). Available mass spectra (e.g., Figure 5.8) do not show ions below $m/z$ 30, so that no information is available about the intensities of any ions that might be present at $m/z$ 15 and 1.

Clearly the fragmentation process leading to formation of $m/z$ 44 has the lowest $\Delta G^{\ddagger}$. Why? Since, with the exception of the broken bonds, the arrangements of atoms in the product ions are the same as those in the molecular ion, it seems safe to assume entropy factors play a relatively minor role in determining values for $\Delta G^{\ddagger}$. Instead, differences in $\Delta G^{\ddagger}$ are reflected in the enthalpy factors that help stabilize the transition states for these reactions. In this case, these appear to be the same factors that contribute to relative stabilization of the products themselves.

**Table 3.1. α-Cleavages of Amphetamine**

| Charge Retention | | Charge Migration | |
|---|---|---|---|
| Loss of H· | ($m/z$ 134; (<1%) | Formation of $H^{+}$ | ($m/z$ 1) |
| Loss of $CH_3$· | ($m/z$ 120; 1–2%) | Formation of $CH_3^{+}$ | ($m/z$ 15) |
| Loss of $\phi CH_2$· | ($m/z$ 44; 100%) | Formation of $\phi CH_2^{+}$ | ($m/z$ 91; 5–10%) |

Consider first the charge retention mechanisms. In each case, the ion product has the positive charge stabilized by participation of the unpaired $n$-electron on the nitrogen. The relative inductive effects on each of these ions—or, more accurately, on the *incipient ions* in the transition state—by the aliphatic groups attached to the doubly bonded carbon should be fairly similar (the aromatic ring in the benzyl group, $\phi$-$CH_2$, is isolated from this carbon and cannot participate in extended conjugation with the $C=N$ bond). The incipient radicals, on the other hand, have strikingly different stabilities, with methyl somewhat more stable than hydrogen because of inductive effects, but substantially less stable than benzyl, which is resonance-stabilized. On this basis, the transition state for the mechanism involving benzyl radical formation is stabilized (has the lowest $\Delta G^{\ddagger}$) relative to the others by the resonance energy of the incipient benzyl radical. Ions formed by this pathway will be formed more rapidly, and thus in greater abundance, than ions formed by the other pathways.

Another factor appears to be at work in the charge migration mechanism. Since benzyl ion, being an "aromatic" ion (see Figure 4.7), should be at least as stable as any of the ions formed by charge retention, and since the radical product formed along with benzyl ion has no obviously inherent instability, it is surprising that the ion resulting from this fragmentation ($m/z$ 91) is so small in the spectrum. One must conclude that just moving the charge from one part of the molecule to another substantially destabilizes the transition state in the charge migration mechanism relative to those involving charge retention. Even the increased stability of the ion product is not sufficient to overcome this destabilization.

## 3.4. PROBLEMS

***Problem 3.1.*** Although no data are given in Table 3.1 for the relative intensities of the $m/z$ 1 and 15 ions in the amphetamine mass spectrum, we expect these ions to be fairly small. Why?

***Problem 3.2.*** Write detailed mechanisms for the following simple fragmentations, showing initial ionization sites and subsequent electron movement:

    a. Formation of ethyl radical and $CH_2=NH_2 \oplus$ from $n$-propylamine
    b. Formation of methyl radical and allyl carbocation ($CH_2=CH-CH_2 \oplus$) from 1-butene
    c. Loss of chlorine radical from isopropyl chloride
    d. Loss of methyl radical from acetone ($CH_3COCH_3$)
    e. Loss of carbon monoxide from cyclohexanone

# CHAPTER 4

# NEUTRAL LOSSES AND ION SERIES

## 4.1. NEUTRAL LOSSES FROM THE MOLECULAR ION

An unknown compound usually cannot be identified from its mass spectrum unless the molecular ion is determined correctly. Figuring out what fragments are lost directly from the molecular ion not only helps confirm the correct choice of the molecular ion, it also provides essential clues about what functional groups are present in the molecule. Table 4.1 lists the most common fragments up to 45 daltons that can be lost as neutral species. You may want to mark this table for easy reference, especially for solving unknowns, until the losses listed become more familiar to you.

Although much of Table 4.1 is self-explanatory, some general observations need to be made. First, this is not an exclusive list; it specifically does not list combinations of losses. Many steroids, for example, show an ion resulting from the loss of 33 daltons from the molecular ion. These compounds also exhibit ions due to the losses of water at $(M - 18)^+$ and methyl at $(M - 15)^+$, so that the ion at $(M - 33)^+$ undoubtedly arises from consecutive losses of methyl *and* water. Still, there is no single fragment of 33 daltons that organic compounds can lose.

Second, this list is useful *only for losses from the molecular ion*. It is very tempting to think that any two ions separated by 15 daltons are related by methyl radical loss from the higher mass ion. The problem with doing this is that, unless the parent for the lower mass ion is known with certainty by some independent means, even an "obvious" situation can lead to erroneous assumptions. For example, the presence of ions at $(M - 15)^+$ and $(M - 30)^+$ may not mean that methyl radical is lost from the $(M - 15)^+$ ion. If the molecule contains oxygen, the $(M - 30)^+$ ion can arise via loss of formaldehyde ($CH_2O$) directly from the molecular ion. The latter possibility is

**Table 4.1. Fragments Commonly Lost as Neutral Species**

| | | | |
|---|---|---|---|
| M-1 | H· | M-29 | $CH_3CH_2$· ; HCO· |
| M-15 | $CH_3$· | M-30 | NO· (nitro cmpds) ; $H_2CO$ (anisoles) |
| M-16 | O· (rare); $NH_2$· (amides) | M-31 | $CH_3O$· |
| M-17 | OH· ; $NH_3$ (rare) | M-32 | $CH_3OH$ |
| M-18 | $H_2O$ | M-35 | Cl· $^a$ |
| M-19 | F· | M-36 | $HCl^a$ |
| M-20 | HF (very rare) | M-42 | $CH_2{=}C{=}O$ ; $CH_2{=}CH{-}CH_3$ |
| M-26 | HCCH, CN | M-43 | $CH_3CO$· ; $C_3H_7$· |
| M-27 | HCN | M-44 | $CO_2$ |
| M-28 | CO ; $CH_2{=}CH_2$ | M-45 | $CH_3CH_2O$· ; ·$CO_2H$ |

$^a$ Check for loss of or change in isotopic abundance pattern.

more useful, as we shall see later, in helping to identify the presence of a phenyl-methylether linkage in the molecule.

Third, there are no common organic fragments between hydrogen and methyl that can be lost. This is not to say that no ions are ever observed between $(M - 1)^+$ and $(M - 15)^+$. With aromatic compounds, for example, it is not uncommon to see spectra that exhibit losses of several hydrogens. In such cases, however, ions corresponding to each of these consecutive losses are observed. An $(M - 6)^+$ ion, without the intervening $(M - 1)^+$, $(M - 2)^+$, and so forth ions being prominent as well, should be viewed suspiciously. An extremely important corollary to this observation is that **loss of a 14-dalton fragment is never observed**. Because of the inherent instability of atomic nitrogen (N) and methylene ($CH_2$), these groups are never lost from the molecular ion or from any other ion for that matter.

In addition to the gap between 1 and 15 daltons, there are three other gaps in the table—between 20 and 26, 32 and 35, and 36 and 42 daltons—that provide tools for evaluating the choice for molecular ion in an unknown spectrum. Although consecutive losses may account for ions at $(M - 33)^+$, $(M - 34)^+$ or $(M - 37$ to $41)^+$ daltons under some circumstances, ions in these areas *not* accompanied by relevant smaller losses should be viewed suspiciously. Indeed the presence of ions in these areas of the spectrum usually indicate either that the spectrum is not that of a pure compound or that the postulated molecular ion is not, in fact, the molecular ion.

Some losses listed in Table 4.1 are very common and will become familiar as we discuss spectra in more detail. Most notable among these are the losses of hydrogen radical, methyl radical, water (from certain oxygenated compounds), acetylene (HC≡CH; from aromatic compounds), hydrogen cyanide (HC≡N; from aromatic compounds containing nitrogen in or on the ring), both carbon monoxide (C≡O) and ethylene ($CH_2{=}CH_2$) as 28 daltons (making it difficult to tell which has been lost!), ethyl radical, methoxy radical ($CH_3O$·), chlorine radical and HCl from chlorinated compounds, and both acetyl ($CH_3CO$·; usually accompanied by a prominent ion at $m/z$ 43 for the acetyl ion) and propyl radicals as 43 daltons.

Other losses are rarer or occur only with specific types of functional groups. For example, the loss of 16 daltons occurs infrequently and is nearly specific either for primary amides (where it is lost as $NH_2\cdot$) or for certain formally polar nitrogen–oxygen bonds such as those found in nitrogen oxides and nitro groups (where the loss is atomic oxygen with its *six* valence electrons). Also, although the loss of water from oxygen-containing compounds is commonplace, the corresponding loss of ammonia from amines is rare. Thus, loss of 17 daltons is assumed to be that of hydroxyl radical unless independent evidence dictates otherwise.

## 4.2. OVERALL APPEARANCE OF THE SPECTRUM— LOW MASS ION SERIES

The mass spectra of saturated aliphatic hydrocarbons (e.g., *n*-decane; Figure 4.1) are visually quite different from those of the condensed polynuclear (multiring) aromatic compounds shown in Figure 4.2. The aliphatic hydrocarbon spectrum exhibits an abundance of low mass ions, making it look as if the molecule has a tendency to just "fall apart." The aromatic compounds, on the other hand, show very little fragmentation; in each case the molecular ion is also the base peak in the spectrum. These features, along with isotopic abundance patterns, can offer insight into the type of compound giving rise to the spectrum.

### 4.2.1. Aliphatic Ion Series

The spectrum of *n*-decane is typical of the spectra of unbranched alkanes. All of these compounds produce a molecular ion, although the relative abundance of this ion decreases with increasing chain length. They show little or no loss of methyl, then appear to progressively lose alkyl groups in a pattern in which the most intense ions in each group are separated by 14 daltons, with peaks of lesser intensity 1 or 2 daltons below the main ion in each cluster. The base peak in these spectra is often *m/z* 43, although occasionally *m/z* 57 proves to be more intense.

At first it might seem strange that these compounds lose alkyl groups of virtually every size except methyl [in Figure 4.1 the major losses appear to be $(M - 29)^+$, corresponding to loss of ethyl radical, $(M - 43)^+$ from loss of a propyl group, $(M -$

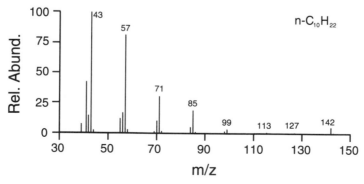

**Figure 4.1.** Mass spectrum of a straight-chain hydrocarbon (*n*-decane).

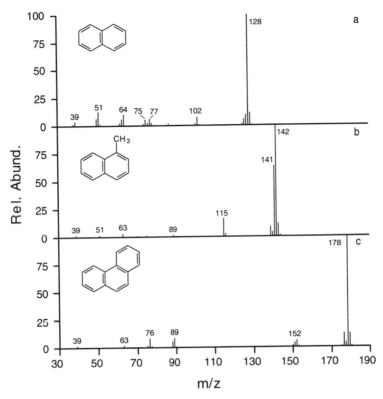

**Figure 4.2.** Mass spectra of three polynuclear aromatic compounds: (*a*) naphthalene, (*b*) 1-methylnaphthalene, and (*c*) phenanthrene.

57)$^+$ from loss of a butyl fragment, etc.]. There are two reasons for this apparent behavior. First, because all of the carbon–carbon bonds in these molecules are of similar strength, initial ionization can occur virtually anywhere along the chain. Thus, all primary fragmentations of these compounds involve the formation of an alkyl radical and a primary carbocation:

$$R^1\text{-}CH_2\text{-}CH_2 \vdots CH_2\text{-}R^2 \longrightarrow R^1\text{-}CH_2\text{-}CH_2^{\oplus} + \ ^{\bullet}CH_2\text{-}R^2 \qquad (4.1)$$

$$\text{OE} \overset{+}{\cdot} \qquad\qquad\qquad \text{EE}^{\oplus} \qquad\qquad \text{OE}^{\bullet}$$

Since all of the primary carbocations are similar in stability, the relative stability of the radicals formed determines which fragments are lost (Section 3.3). The stability of these radicals follows the pattern:

Hydrogen < methyl < ethyl < *n*-propyl < isopropyl < *t*-butyl

with no further stability being gained by increasing the length of the carbon chain.

Isopropyl and *t*-butyl radicals are more stable because they are, respectively, secondary and tertiary radicals; the stability of the listed primary radicals increases somewhat with the added inductive effect of larger alkyl substituents.

The second reason for lack of methyl loss is obscured by the fact that the mass spectra, by appearing to depict the loss of larger and larger alkyl groups, do not adequately reflect the complexity of the actual fragmentation processes. In particular, since primary radicals larger than *n*-propyl gain no further stability from increasing chain length (even the energy gain from ethyl to propyl is slight), the energies of the transition states resulting from the incipient product mixtures of primary carbocations and radicals are essentially the same whether these compounds lose ethyl, propyl, butyl, or pentyl radicals. This seems to contradict the fact that the peaks at lower mass are more intense; however, this is not the only fragmentation occurring! Instead, additional studies have shown that the lower mass ions result from secondary (and further) fragmentations in which the positive charge is relocated on a site of similar stability (all of the ions formed are primary carbocations), and in which a *more* stable neutral product is formed—in this case an olefin (ethylene; $CH_2{=}CH_2$) formed by rearrangement of the electron density:

$$R^1\text{-}CH_2\text{-}CH_2\text{-}CH_2^{\oplus} \longrightarrow R^1\text{-}CH_2^{\oplus} + CH_2{=}CH_2 \qquad (4.2)$$

$$EE^{\oplus} \qquad\qquad EE^{\oplus} \qquad\qquad EE^{\circ}$$

Thus, although the ion at *m/z* 113 in Figure 4.1 is indeed formed by loss of ethyl radical from the molecular ion, the ion at *m/z* 85 arises from a combination of butyl radical loss *plus* loss of ethylene from the ion at *m/z* 113. Similarly, the ion at *m/z* 57 can form by loss of ethylene from the *m/z* 85 ion. The driving force for this fragmentation is formation of the new π-bond in the extruded molecule of ethylene (Section 3.3), which lowers the activation energy toward secondary fragmentation of the high mass ions. This is the first of many examples in which loss of a small, multiply bonded molecule relocates the positive charge to a site that has stability similar to that seen in the parent ion.

The series of intense ions at *m/z* 43, 57, 71, 85, 99, and so forth is highly characteristic of saturated aliphatic hydrocarbons in general (see also Figures 4.13 and 4.14). A group of ions at low mass that appears consistently in the spectra of a class of compounds is called a *low mass ion series,* and constitutes, after the characteristic isotopic abundance patterns discussed Chapter 2, one of the first things to look for in an unknown spectrum for clues about the nature of the compound under consideration (refer back to Section 2.3).

Aliphatic compounds bearing functional groups often have their own distinctive low mass ion series that, in addition to those for some aromatic compounds to be discussed below, are listed in Table 4.2. Each ion series listed has some ions highlighted in boldface type—these ions tend to be more intense than other ions *in that area of the spectrum;* it does not necessarily mean that they are the most intense ions in the spectrum or even that they are intense ions.

Inserting one or more double bonds into an aliphatic compound, or cyclization into a saturated ring system, leads to spectra in which, like those of saturated aliphatics,

**Table 4.2. Common Ion Series at Low Masses**

1. General (of little or no value in determining specific structural features).
    a. Saturated aliphatics $[(C_nH_{2n+1})^+]$:
       27, **29**; 41, **43**; 55, **57**; 69, **71**; 83, **85**; . . .
    b. Unsaturated aliphatics and cycloalkanes $[(C_nH_{2n-1})^+$ plus rearrangement ions $(C_nH_{2n})^{\ddot{+}}]$:
       27, 29; **41**, 43; **55**, (**56**), 57; **69**, (**70**), 71; **83**, (**84**) . . .
2. α-Cleavage (may be highly specific; one peak usually dominates spectrum).
    a. Ketones $(R-C\equiv O^+)$; do not show full aliphatic pattern:
       **43**; **57**; **71**; **85**; . . .
    b. Ethers and alcohols $(R_2C=OR^+)$:
       **31**; **45**; **59**; **73**; **87**; . . .
    c. Amines $(R_2C=NR_2^+)$:
       **30**; **44**; **58**; **72**; **86**; **100**; . . .
3. Aromatic (usually gives general information only; may be specific for benzyl and benzoyl).
    a. With electron-donating substituents (alkylbenzenes, ethers, etc.):
       **39**; 50, **51**, 52; 63, 64, **65**; 76, **77**, 78
    b. With electron-withdrawing substituents (nitro, halogens, etc.):
       **38**, **39**; 49, **50**, 51; 62, **63**, 64; **75**, **76**, 77
    c. Benzyl $(\phi CH_2^+$; may be specific if $m/z$ 91 is very intense):
       39, **65**, **91**
    d. Benzoyl $(\phi CO^+$; may be specific if $m/z$ 105 is intense):
       **51**, **77**, **105**

low mass ions tend to dominate (Figure 4.3). Notice, however, that the most intense ions in each cluster are not the same as those for saturated aliphatic compounds; most are found 2 daltons lower, reflecting stabilization of the positive charge by the unsaturation (see Chapter 5). Thus, $m/z$ 41 corresponds to the allyl carbocation ($CH_2$=CH-$CH_2\oplus \leftrightarrow \oplus CH_2$-CH=$CH_2$), $m/z$ 55 to the methylallyl ion, and so forth. The ions at $m/z$ 56 in the top and bottom spectra, and $m/z$ 70 in all three spectra, are odd-electron ions (Section 3.1), resulting from rearrangement.

The addition of electron-withdrawing functional groups, such as carbonyl, ether, alcohol, or amine groups, to a saturated aliphatic compound drastically changes the appearance of the spectrum (Figure 4.4). These groups so highly polarize a few bonds in these molecules that fragmentation is centered almost entirely on them, in many cases producing one or two ions of unusual intensity and stability. This type of fragmentation, called α-cleavage, is one of the most distinctive and important we will discuss. The mass spectra of these compounds will be covered in detail in Chapter 5.

Distinctive ion series characterize other specific types of compounds and become obvious after the spectra of several examples from the family have been studied and compared. For example, terpenes are isomeric unsaturated cyclic hydrocarbons found in various products derived from the distillation of wood and are frequently encountered in the forensic analysis of arson residues. The spectra of the three terpenes shown in Figure 4.5 have a surprising number of similarities, despite substantial differences in structure. The ions at $m/z$ 77, 79, 93, 107, 121, and 136 prove to be widespread among this family of compounds.

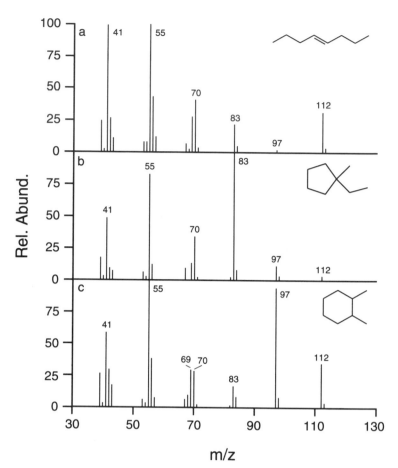

**Figure 4.3.** Mass spectra of olefinic and alicyclic compounds: (*a*) 4-octene; (*b*) 1-methylethyl-cyclopentane; and (*c*) 1,2-dimethylcyclohexane. Note the odd-electron ions at *m/z* 70.

### 4.2.2. Aromatic Ion Series

In contrast to the alkane spectrum in Figure 4.1, the spectra of the aromatic compounds in Figure 4.2 lack an intense low mass ion series. Instead, low mass peaks in these spectra are small and vary in relative intensity from spectrum to spectrum. The intensities of these fragment ions reflect their relative instability compared to that of the aromatic molecular ions and the high energies of activation that must be overcome to force fragmentation to occur. Despite the fact that the ion series for aromatic compounds is not as well defined as that for saturated aliphatics, it is nonetheless consistent enough to be characteristic of these compounds.

In the spectrum in Figure 4.2*a*, the aromatic low mass ion series consists of the ions at *m/z* 38, **39**; 50, **51**; 62, 63, **64**; and 74, **75**, 76, **77**, 78. (We might also include the ion at *m/z* 89 in the ion series for these condensed polynuclear aromatic com-

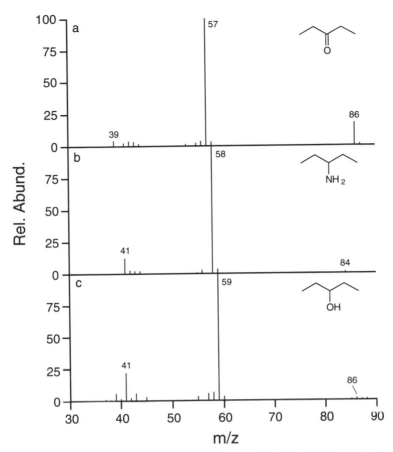

**Figure 4.4.** Mass spectra of aliphatic compounds with functional groups: (*a*) 3-pentanone, (*b*) 3-pentylamine, and (*c*) 3-pentanol.

pounds, but some substituted benzenes do not show ions in this area of the spectrum, so it is not listed in Table 4.2.) This ion series matches well with that listed in Table 4.2 for aromatic compounds having electron-withdrawing groups (series 3b), which accurately describes naphthalene, in which one aromatic ring withdraws electron density from the other.

This series differs slightly from that for aromatic compounds having electron-donating subtituents (series 3a). In this case the more intense ions in each group tend to occur 1 to 2 daltons higher in mass (Figure 4.6). This difference can be rationalized if we consider likely structures for these ions (Figure 4.7). Whereas the ions at $m/z$ 91, 77, 65, 51, and 39 (typical of compounds with electron-donating substituents) show the degree of unsaturation expected for aromatic compounds, those at $m/z$ 89, 76, 63, and 50 have structures with triple bonds in the aromatic rings ("benzyne-type ions"). The ion at $m/z$ 76 is the molecular ion of benzyne ($C_6H_4$), a reactive interme-

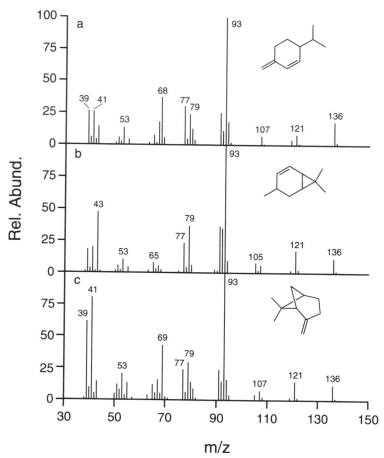

**Figure 4.5.** Mass spectra of many terpenes are very similar, despite substantial differences in structure: (*a*) 3-methylene-6-isopropylcyclohexexe (menthadiene); (*b*) 4,7,7-trimethylbicy-clo[4.1.0]hept-2-ene (4-carene); and (*c*) 2-methylene-6,6-dimethylbicyclo[3.1.1]heptane (ß-pinene).

diate known from solution chemistry that can be trapped by reaction with suitable reagents. Benzyne is formed most easily in solution by the reactions of benzene derivatives having electron-withdrawing substituents, so that the formation of analogous ions from similar types of compounds in mass spectrometry is not without precedent.

As shown in Figure 4.7, the ions at *m/z* 65, 63, 51, 50, and 39 all form by loss of acetylene from an aromatic ion 26 daltons higher in mass. The loss of small molecules, particularly acetylene and HCN (Section 4.3), from aromatic ions through rearrangement of their π-electronic structure is highly characteristic of such ions.

On the other hand, no simple mechanism can account for formation of the small *m/z* 63 ion in the spectrum of benzene (Figure 4.6*a*). After ruling out the presence of

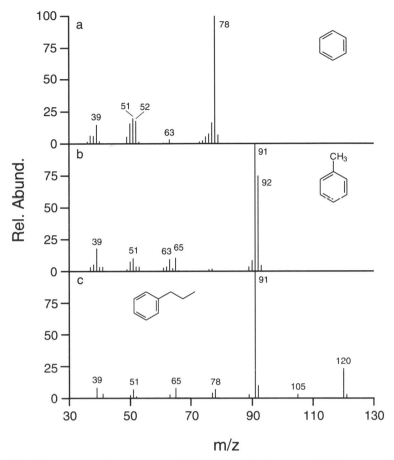

**Figure 4.6.** Mass spectra of some simple alkylbenzenes: (*a*) benzene, (*b*) toluene, and (*c*) *n*-propylbenzene.

an impurity in the spectrum, it becomes extremely difficult to rationalize the contortions through which the benzene molecular ion must go in order to lose a methyl radical. This obviously complex fragmentation reflects again the stability of benzene molecular ion. All of the fragmentation pathways open to it apparently require a substantial input of energy, so that even highly unusual fragmentations compete. As we will see, other aromatic compounds also lose a methyl group from within the ring itself.

Figure 4.7 also illustrates two sets of fragmentations that are so characteristic as to be classified as separate ion series in Table 4.2. The first of these, at the top of Figure 4.7, is the fragmentation of benzyl ion ($C_7H_7^+$; *m/z* 91) by acetylene loss to give the cyclopentadienyl ion at *m/z* 65, which in turn loses acetylene to give the cyclopropenium ion at *m/z* 39. As illustrated, the structure of the benzyl ion is not static. Rather, it is in equilibrium with the cycloheptatrienyl (tropylium) ion and probably with other struc-

Benzyl:

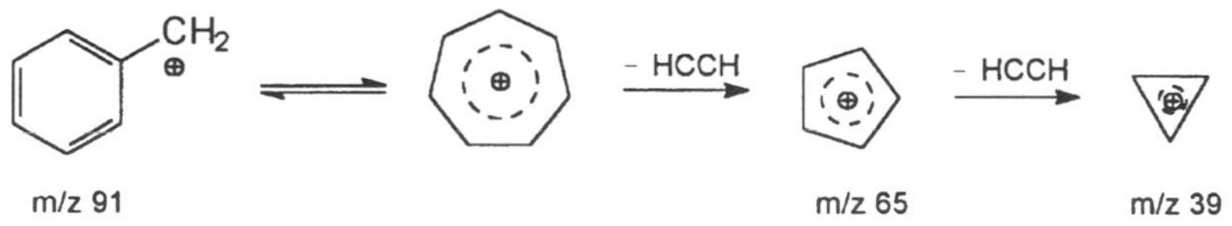
CH2
- HCCH
- HCCH
m/z 91
m/z 65
m/z 39

Benzyne-type ions:

- HCCH
HCCH
m/z 89
m/z 63
m/z 76
m/z 50

Benzoyl:

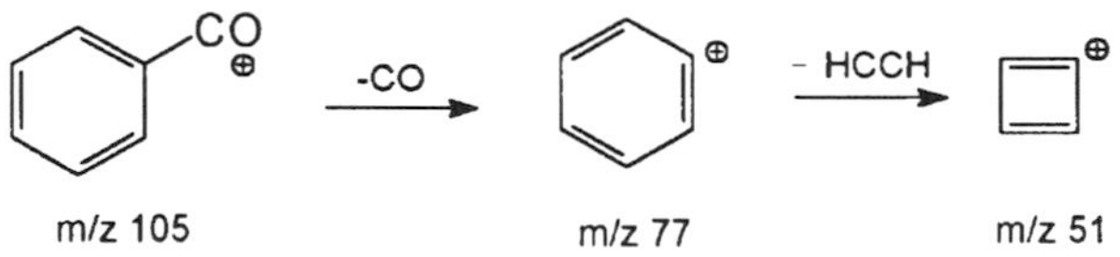
CO
-CO
- HCCH
m/z 105
m/z 77
m/z 51

**Figure 4.7.** Structures and fragmentations of prominent low mass aromatic ions.

tures as well. The cyclic $C_7H_7^+$ and $C_3H_3^+$ structures are examples of nonbenzenoid aromatic systems—that is, cyclic systems of contiguously overlapping *p*-orbitals containing $(4n + 2)$ $\pi$-electrons (where $n = 0,1,2 \ldots$; Hückel's rule) that do not have the traditional benzene structure. With benzene and the tropylium ion, $n = 1$ and $4(1) + 2 = 6$, whereas for the cyclopropenium ion, $n = 0$ and $4(0) + 2 = 2$. Both the tropylium and the cyclopropenium ions are so stable, in fact, that their salts can be isolated.

The benzyl ion series is seen clearly in the spectra of toluene and propylbenzene in Figure 4.6—these fragmentations are predictable and characteristic for both of these compounds (Chapter 5). However, because of the stability of the benzyl/tropylium ion, even aromatic compounds lacking a $\phi$-$CH_2$ group may fragment with hydrogen rearrangement just so that the benzyl/tropylium ion can be formed. A good example is the hallucinogenic drug phencyclidine (structure below), whose mass spectrum shows

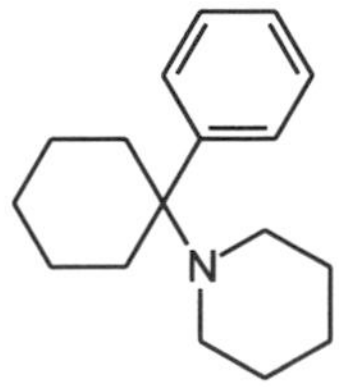
N

$m/z$ 91 as the most abundant low mass ion (25 to 50% relative intensity) even though two hydrogens must be rearranged to the benzylic carbon in its formation. Thus, unless $m/z$ 91 is very intense, the presence of a benzyl ion series may be more an indicator of aromaticity than specifically of a benzyl group in the molecule.

The benzoyl series (Figure 4.7, bottom) is more reliable. An aromatic ring with an attached carbonyl group—whether as a ketone, ester, amide, and so forth—shows a strong propensity to fragment in such a way that benzoyl ion ($C_6H_5CO^+$ ; $m/z$ 105) is a prominent ion in the spectrum. Indeed, for simple aromatic carbonyl compounds, the benzoyl ion series may account for nearly all of the most intense ions in the spectrum (see Figures 5.18 to 5.20). A subtler example is seen in the spectrum of the stimulant drug cocaine (Figure 8.4). As with other aromatic ions, the benzoyl ion fragments through the loss of small, triply bonded molecules—first carbon monoxide, then acetylene.

### 4.2.3. Uses of Ion Series—Reconstructed Ion Chromatography

In addition to offering visual cues about the type of compound giving rise to a mass spectrum, low mass ion series can be used to locate families of compounds in chromatograms of complex mixtures. An excellent example of this is found in the forensic analysis of arson residues.

Figure 4.8 shows total ion chromatograms (TICs) for some typical arson accelerants. These TICs were obtained by collecting full-scan mass spectra continuously throughout a GC/MS run, then plotting the total current output of the electron multiplier for each scan vs. time (see Section 1.5.4). The resulting chromatogram appears similar to those produced by other GC detectors, but the stored mass spectral data has an additional wealth of information that can be used for compound identification.

For each of the mixtures in Figure 4.8, the chromatographic pattern itself is complex and reproducible enough that it constitutes a characteristic feature for identification purposes even without knowledge of the individual components. Unfortunately, the chromatograms produced by evidence collected from the scenes of suspect fires may bear only a vague resemblance to those of standards because of the extreme conditions to which the samples are exposed. It has proven virtually impossible to reproduce the combination of evaporation, oxidation, weathering, and, in some cases, even biological degradation that accelerants undergo during an arson fire, so that meaningful "standard" chromatograms are not available. In addition, the pyrolysis of both natural and synthetic materials present at the scene produces mixtures of volatile products that can be confused with accelerants in some cases (Figure 4.9).

For these reasons arson residues must be characterized by identification of their constituents. Even this proves problematic since many commercial products commonly used as arson accelerants are petroleum distillates and thus contain many of the same compounds, albeit in different proportions. As a result, identifying a certain number of individual components by mass spectrometry may not identify the accelerant at all.

Instead of "absolute" identification using the entire mass spectrum, many compounds can be "characterized" (identified with significant, but somewhat less, cer-

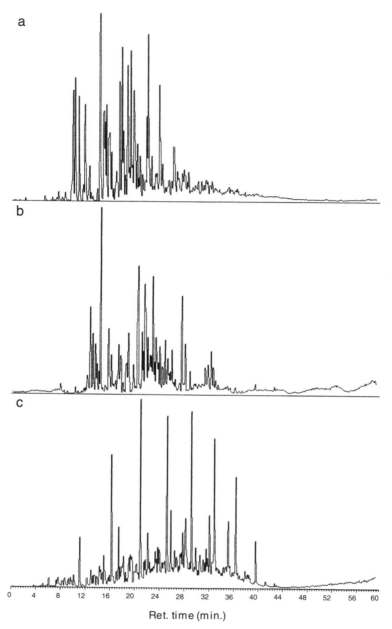

**Figure 4.8.** Mass spectrometer as GC detector—total ion chromatograms of three volatile liquids. (*a*) Charcoal lighter fluid; (*b*) evaporated gasoline; (*c*) fuel oil.

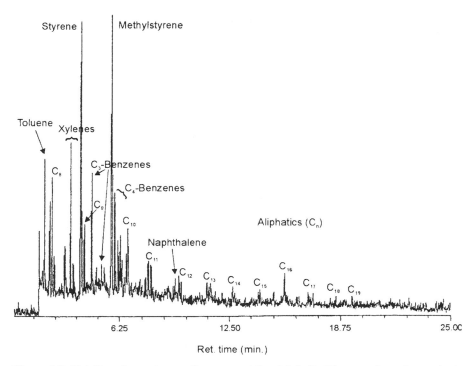

**Figure 4.9.** Total ion chromatogram from suspect fire debris (in this case, charred carpeting). Compounds formed by pyrolysis of common polymers are often the same as those found in arson accelerants.

tainty) by a combination of their GC retention times and a few critical ions in their spectra. Low mass ion series offer a way to provide this type of characterization. For example, saturated aliphatic hydrocarbons produce characteristic intense ions at $m/z$ 43, 57, 71, and 85, while unsaturated and alicylic compounds show peaks at $m/z$ 41, 55, 69, and 83 (Table 4.2). Although aromatic compounds might be characterized by any of the appropriate ion series in Table 4.2, the aromatic compounds actually present in petroleum distillates can be broken down further into separate families, each with their own distinctive set of ions (Table 4.3).

These data are retrieved by the computer as reconstructed ion chromatograms (RICs), in which the multiplier ion current for individual ions or small group of ions are plotted vs. time (Section 1.5.4). Reconstructed ion chromatograms are produced from existing GC/MS data and can be generated after the sample has been run for any ion in the scan range. Selected ion monitoring (Section 1.3.2), on the other hand, produces similar information only for ions preselected before the run begins. Although using reconstructed ion chromatography appears to enhance sensitivity because it can help locate miniscule peaks buried in complex chromatograms, only selected ion monitoring actually lowers the limits of detection due to the increased amounts of time the analyzer spends monitoring the masses of specific ions.

**Table 4.3. Aromatic Ions Characteristic of Petroleum Distillate Constituents**

| Type of Compound | Characteristic Ions |
|---|---|
| Alkylbenzenes | 91, 105, 119, 133 (fragment ions) |
| | 78, 92, 106, 120, 134, etc. (molecular ions) |
| Alkylnaphthalenes | 128, 142, 156, 170 (molecular ions) |
| Alkylstyrenes and indenes | 104, 118, 132, 146 |
| Alkylanthracenes and phenanthrenes | 178, 192, 206 |

Some data systems are capable not only of searching mass spectral data for the ion current from individual ions but also of adding the responses for several ions at a time, producing a "family ion chromatogram." Summed reconstructed ion chromatograms for some of the hydrocarbon families found in evaporated gasoline are shown in Figure 4.10. Most striking is the pattern of peaks shown in the "naphthalenes" chromatogram, where the individual homologs are easily discernible. With experience the trained analyst can get appreciable insight into the composition of a sample based on this type of information.

## 4.3. OTHER SMALL MOLECULE LOSSES IN AROMATIC COMPOUNDS

In the spectrum of pyridine (Figure 4.11a) the electron-donating aromatic low mass ion series is truncated, showing only the ions at $m/z$ 39 and 50 to 52. Other than the sequential loss of several hydrogen radicals, the first major loss from the molecular ion is 27 daltons. Although this loss could occur in several different ways, a mechanism involving the four-center loss of hydrogen cyanide parallels the loss of acetylene in aromatic compounds that do not contain nitrogen:

$$\tag{4.3}$$

m/z 79     m/z 52

The propensity for compounds containing adjacent nitrogens to lose molecular nitrogen is well-known from ground-state organic chemistry. It should not be surprising, then, that these electronically similar molecules should be lost so easily from highly energetic ions:

$$H{:}C{\equiv}C{:}H$$

$$H{:}C{\equiv}N{:}$$

$${:}N{\equiv}N{:}$$

$$^-{:}C{\equiv}O{:}^+$$

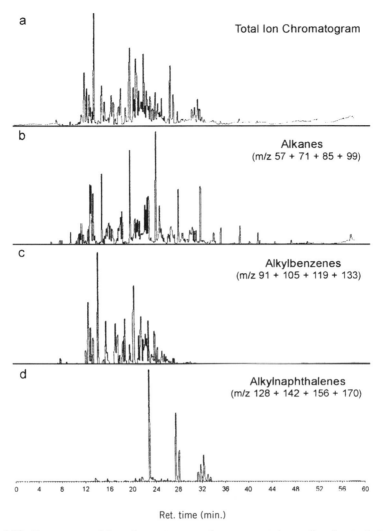

**Figure 4.10.** Reconstructed ion chromatograms for evaporated gasoline locate individual components in complex mixtures. Summed chromatograms here are based on common low mass ion series.

One additional comment needs to be made about the spectrum of pyridine. The molecular weight of this compound is odd, which contrasts with those of most of the compounds discussed thus far. Of all of the common elements (C, H, O, S, halogens, etc.), nitrogen is the only element that, in organic compounds at least, has an odd valence (3) and an even atomic weight (14). This combination results, as noted previously (Section 2.3), in the *odd-nitrogen rule:* If an organic compound contains an odd number of nitrogen atoms (1, 3, 5, . . . ), it will have an odd molecular weight; if the compound contains an even number of nitrogen atoms (0, 2, 4, . . . ), it will have an

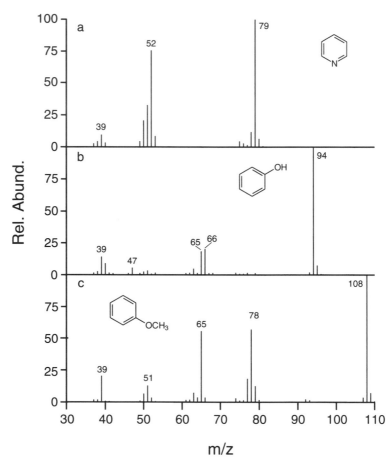

**Figure 4.11.** Mass spectra of aromatic compounds containing heteroadams: (*a*) pyridine, (*b*) phenol, and (*c*) anisole.

even molecular weight. There are essentially no exceptions to this rule. The real utility of the odd-nitrogen rule is in its converse—namely, that if a compound has an odd molecular weight, it *must* contain an odd number of nitrogens, and if it has an even molecular weight, it *must* contain an even number of nitrogens. Although this rule holds for *all* odd-electron ions (of which the molecular ion is one), it is almost impossible to tell whether any given ion in the middle of the spectrum of a nitrogen-containing compound is odd-electron or not. Thus use of this rule should be limited to the molecular ion alone.[1]

The spectrum of phenol (Figure 4.11*b*) exhibits a typical aromatic low mass ion series at *m/z* 39, 51, and 65, but two other ions in this spectrum deserve comment. The first of these, at *m/z* 47, has an unusual mass in that there are few combinations of elements in organic compounds that can produce ions of this mass. In fact, the range from *m/z* 46 through 49 is generally devoid of peaks with the exception of compounds

---

[1] The spectrum of 3-pentylamine shown in Figure 4.4*b* does not exhibit a molecular ion. The peak at *m/z* 84 is the $(M - 1)^+$ ion.

containing sulfur ($CH_3S^+$ at $m/z$ 47) and chlorine ($CCl^+$ at $m/z$ 47 and 49). In the case of phenol, there is no reasonable combination of elements that can account for this ion. However, the molecular weight for this compound (94) is exactly twice 47. Thus the ion at $m/z$ 47 can be ascribed to the doubly charged molecular ion ($C_6H_6O^{2+}$). Since the mass spectrometer measures the mass-to-charge ratio, not just the mass, of each ion, doubly charged ions appear at M/2, or half the mass at which we expect to see them. Doubly charged ions, which are usually weak, are not commonly identified in low-resolution spectra. They usually occur only in the spectra of aromatic compounds, which often have few fragmentation modes open to them, or compounds containing heteroatoms and having low second ionization potentials.

The other ion of note in this spectrum is at $m/z$ 66, involving the loss of 28 daltons from the molecular ion. From the list in Table 4.1, we see that 28 daltons can be lost either as carbon monoxide or as ethylene; both losses occur frequently. However, loss of ethylene from an aromatic ring would involve substantial hydrogen and electronic rearrangement, in contrast to previous examples where two-carbon fragments in aromatic compounds were lost as acetylene. Although loss of carbon monoxide must be accompanied by hydrogen rearrangement as well, one plausible mechanism has precedent in solution chemistry.

This fragmentation can be rationalized in the following way:

(4.4)

First, interconversion between the *enol* form of phenol and its tautomeric *keto* form via a four-center hydrogen migration removes the hydrogen from the oxygen. Although this equilibrium lies far to the left (>99% enol) due to the resonance energy of the aromatic ring, any reaction that disturbs the equilibrium by removing the keto form (such as the proposed fragmentation), will cause further formation of the keto form (LeChatlier's principle).

The second step involves breaking one of the carbon–carbon bonds attached to the carbonyl group, stabilizing the positive charge on both the carbonyl carbon and the carbonyl oxygen and forming an allylic radical ($\alpha$-cleavage; see Chapter 5). The double bond nearest the carbonyl group then donates one electron to form a new C—C single bond, close the five-membered ring, and generate an equally stable allylic radical site—all the while retaining the positive charge on the carbonyl oxygen. Finally, cleavage of the C—C bond between the ring and the carbonyl group, with movement of the electron pair onto the carbonyl carbon, expels a neutral molecule of carbon

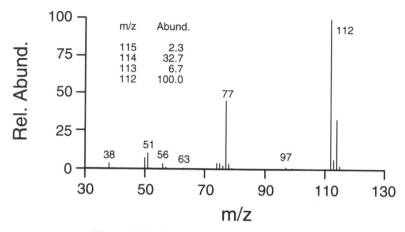

| m/z | Abund. |
|-----|--------|
| 115 | 2.3 |
| 114 | 32.7 |
| 113 | 6.7 |
| 112 | 100.0 |

**Figure 4.12.** Mass spectrum for Problem 4.1.

monoxide and forms the molecular ion of cyclopentadiene [$m/z$ 66; Eq. (4.4)]. The ion at $m/z$ 65 forms by hydrogen radical loss from $m/z$ 66, while the ions at $m/z$ 40 and 39 result from loss of acetylene (26 daltons) from the $m/z$ 66 and 65 ions, respectively.

The spectrum in Figure 4.11c is that of anisole (phenyl methyl ether). What is striking about this spectrum is the intense ion at $m/z$ 78, corresponding to the loss of 30 daltons from the molecular ion. Studies have shown that the $m/z$ 78 ion is the benzene molecular ion, so that the fragment lost is formaldehyde ($CH_2O$). As in the previous example, this loss involves both a preliminary four-center hydrogen migration and subsequent loss of a small multiply bonded molecule, shown schematically as:

$$\tag{4.5}$$

m/z 108                                                  m/z 78

The loss of formaldehyde in this manner is characteristic of anisoles in general.

**Problem 4.1.** Identify the compound whose spectrum is shown in Figure 4.12.

## 4.4. EFFECT OF CHAIN BRANCHING IN THE SPECTRA OF ALIPHATIC HYDROCARBONS

Although the spectra of saturated aliphatic hydrocarbons have many features in common, they contain enough differences to permit structure identification based on their fragmentation patterns. At the same time it is important to realize that, for compounds containing more than seven or eight carbons, the large number of isomeric possibilities often precludes making a one-to-one assignment of structures to spectra. The following discussion is intended to be helpful, rather than definitive.

The spectra of the three octanes in Figure 4.13 illustrate the differences caused by increasing the amount of branching. Whereas the spectrum of *n*-octane (Figure 4.13*a*) shows a typical straight-chain hydrocarbon pattern, that of 2-methylheptane (Figure 4.13*b*) has a sizable $(M - 15)^+$ ion. Although *n*-alkanes do not lose methyl radical because larger radicals are lost preferentially in otherwise energetically similar processes [Eq. (4.1)], 2-methylheptane forms a *secondary* carbocation with methyl loss, lowering the activation energy for this fragmentation over those in which primary carbocations are formed. Loss of pentyl radical from the molecular ion also leads to a secondary carbocation at *m/z* 43, and, since pentyl radical is more stable than methyl radical, we might expect to see an enhancement in the size of *m/z* 43 relative to other ions. However, the formation of low mass ions by the complex fragmentations shown in Eq. (4.2) obscures this effect, underscoring the difficulty of interpreting the effect of structure on the intensities of low mass ions.

Figure 4.13*c* shows the spectrum of 2,2,3-trimethylpentane, one of the most highly branched octane isomers. In this case *m/z* 57 is not only the base peak, it is also much larger than every other ion in the spectrum except *m/z* 56. A careful look at the structure should indicate why this is so. The bond between carbon atoms 2 and 3 must be weakened even in the parent molecule because of steric interferences between the three methyl groups. Initial ionization at this already weakened bond is thus likely, and breaking this bond to form a *secondary*-butyl radical and a *tertiary*-butyl carbocation should be very favorable energetically:

$$
\begin{array}{c}
\text{CH}_3 \\
|\\
\text{C} \\
\text{CH}_3 \diagup |\quad \text{HC} \diagup \text{CH}_2 \diagdown \text{CH}_3 \\
\quad \text{CH}_3\; |\\
\qquad\quad \text{CH}_3
\end{array}
\longrightarrow
\begin{array}{c}
\text{CH}_3 \\
|\\
\text{CH}_3 \diagup \text{C} \oplus \\
|\\
\text{CH}_3
\end{array}
\; + \;
\begin{array}{c}
\bullet\; \text{CH}_2 \\
\text{HC} \diagup \diagdown \text{CH}_3 \\
|\\
\text{CH}_3
\end{array}
\qquad (4.6)
$$

Alternative loss of methyl radical to form the tertiary 2,3-dimethyl-2-pentyl ion (*m/z* 99) does not compete well at all because of the significant difference in stabilities between methyl and *sec*-butyl radicals. The only other fragmentation seen at high mass is loss of ethyl radical to give the secondary 3,3-dimethyl-2-butyl ion at *m/z* 85. Other low mass ions in this spectrum do not follow the usual pattern for saturated aliphatic compounds but instead emphasize hydrogen loss—for example, butylene $(C_4H_8{}^+$ at *m/z* 56) and allyl $(C_3H_5{}^+$ at *m/z* 41).

Ion stability is thus far more important in determining what fragments are formed from saturated alkanes than is radical stability—that is, secondary carbocations will be formed preferentially over primary, and tertiary ions over secondary. If ion stabilities are equal, however, then radical stability becomes the determinative factor, following the sequence given previously in Section 4.2.1.

***Problem 4.2.*** Figure 4.14 shows the spectra of three $C_7H_{16}$ isomers. Assign structures to these spectra. (*Hint:* First draw the nine possible structures and try to predict fragmentation patterns for each structure based on the discussion above. Then look at the spectra to see which ones most closely match the predicted fragmentation patterns.]

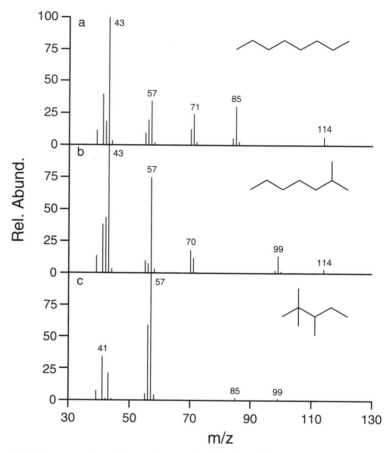

**Figure 4.13.** Mass spectra of three alkanes showing the effects of chain branching: (*a*) *n*-octane, (*b*) 2-methylheptane, and (*c*) 2,2,3-trimethylpentane.

## 4.5. EXAMPLES AND PROBLEMS

***Examples 4.1 to 4.3.*** Identify the compounds whose spectra are shown in Figures 4.15, 4.16, and 4.17.

*Answer 4.1.* This compound has an even molecular weight (100) and none of the obvious heteroatom patterns in any of the major fragment ions. The spectrum appears to have a typical saturated aliphatic low mass ion series at *m/z* 43, 57, and 71, and the molecular weight accommodates an aliphatic hydrocarbon as well—$C_7H_{16}$. The lack of an $(M - 15)^+$ peak is consistent with a straight-chain hydrocarbon since these compounds show little or no loss of a methyl group.

This answer is inconsistent with other aspects of the spectrum, however. First, the isotopic abundances for the ions at *m/z* 57, 77, and 100 show the presence of only 3, 4, and 6 carbons, respectively, not 4, 5, and 7 as expected for a saturated hydrocar-

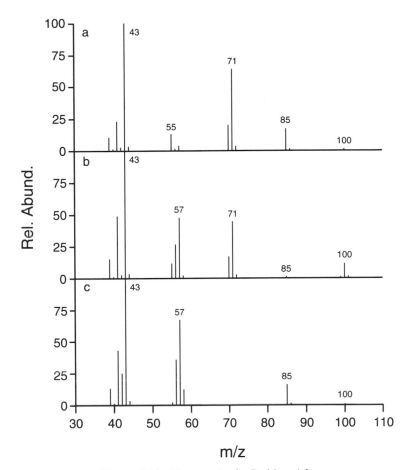

**Figure 4.14.** Mass spectra for Problem 4.2.

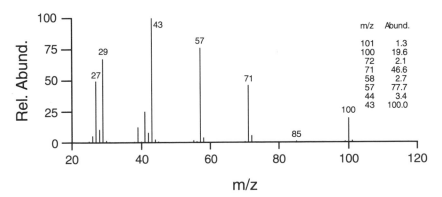

**Figure 4.15.** Mass spectrum for Example 4.1.

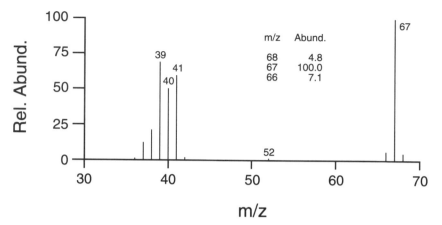

**Figure 4.16.** Mass spectrum for Example 4.2.

bon! Since the presence of nitrogen is not indicated, oxygen must be considered to make up the missing mass. An empirical formula of $C_6H_{10}O$ for the molecular ion does not seem unreasonable as a starting point.

Second, compare this spectrum with that of *n*-decane (Figure 4.1) and *n*-octane (Figure 4.13*a*). Despite the presence of major ions at *m/z* 43, 57, and 71 in all three spectra, the spectra of the *n*-alkanes also contain important ions at *m/z* 70, 56, and 55—ions that are completely absent from the mass spectrum of this unknown. A closer look at Table 4.2 reveals that, in addition to the saturated aliphatic hydrocarbons, aliphatic ketones also have a low mass ion series at *m/z* 43, 57, 71, . . . , except that the full aliphatic pattern of peaks [including the $(P - 1)^+$ and $(P - 2)^+$ ions in each cluster] are not observed. This spectrum is thus that of a 6-carbon aliphatic ketone. Based only on the material covered so far, it is not possible to determine which of several isomers gave rise to this spectrum (3-hexanone; Et = ethyl and Pr = propyl):

$$
\begin{array}{ccccc}
\underset{m/z\ 57}{\overset{\oplus}{O}} & \xleftarrow[a]{-\,Pr^{\bullet}} & \underset{m/z\ 100}{\overset{O^{+\bullet}}{\underset{a}{\text{\Large$\parallel$}}}} & \xrightarrow[b]{-\,Et^{\bullet}} & \underset{m/z\ 71}{\overset{\oplus}{O}} \\
\downarrow{\scriptstyle -\,CO} & & & & \downarrow{\scriptstyle -\,CO} \\
\underset{m/z\ 29}{\overset{\oplus}{\phantom{x}}} & & & & \underset{m/z\ 43}{\overset{\oplus}{\phantom{x}}}
\end{array}
\qquad (4.7)
$$

*Answer 4.2.* The molecular weight of this compound is 67, which tells us immediately that the compound contains an odd number of nitrogens (Section 4.3). Isotopic

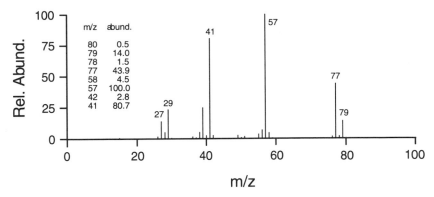

**Figure 4.17.** Mass spectrum for Example 4.3.

abundance information from the $m/z$ 68 ion indicates the presence of four carbons, leading to a molecular formula of $C_4H_5N$. The molecular ion is the base peak, and the only major losses observed are 26 ($HC\equiv CH$ or $\cdot C\equiv N$), 27 (HCN), and 28 daltons (to form $C_3H_3\oplus$). This pattern is consistent with that expected of an aromatic compound. A structure that fits these criteria is pyrrole, which, in addition to the four $\pi$-electrons from the two double bonds, also has the two $n$-electrons from the nitrogen atom to form a cyclic system of overlapping $p$-type orbitals containing six electrons. This conforms with Hückel's rule (Section 4.2.2) for aromatic compounds, and pyrrole in fact exhibits other spectroscopic characteristics and chemical behavior consistent with this definition. The tiny ion at $m/z$ 52 must come from loss of methyl, but this is consistent with a similar loss of a methyl group from benzene (Figure 4.6a) (pyrrole):

*Answer 4.3.* The first step in solving any mass spectral unknown is identifying the molecular ion. Up to this point, some ion in the spectrum of each example or unknown has satisfied the conditions for the molecular ion outlined in Section 2.3. In this spectrum, however, neither of the ions at $m/z$ 77 or 79 can be the molecular ion since, barring severe contamination by some extraneous material, the base peak lies 20 or 22 daltons below them—losses that are forbidden by Table 4.1. Since there are no ions in the spectrum above $m/z$ 80, we cannot identify the molecular ion directly and may only be able to *infer* it from the available information, if we can even accomplish that! Not knowing the molecular weight leaves us in the dark about the presence or absence of nitrogen. Our only hope of solving this problem is to concentrate on what we do know from the spectrum.

The ions at $m/z$ 77 and 79 appear to contain chlorine (the abundance of $m/z$ 79 is about one third that of 77) and three carbon atoms, although we are less sure of these assumptions than in previous cases since we do not know anything about the molecular ion. The ions at $m/z$ 27, 29, 41, and 57 all fall into various hydrocarbon low mass ion series, and the relative abundances of $m/z$ 42 and 58 confirm the presence of three and four carbons, respectively, in $m/z$ 41 and 57. Thus $m/z$ 41 is probably the allyl ion ($\oplus CH_2$-$CH$=$CH_2$), and $m/z$ 57 one of the saturated butyl ions ($C_4H_9^+$). Since there appear to be no hydrocarbon ions above $m/z$ 57 (the next member of the low mass ion series would be at $m/z$ 71) and $m/z$ 77 seems to contain a chlorine and three carbon atoms, it is conceivable that the compound consists of a butyl group and a chlorine. This would lead to a molecular weight of 92 daltons, and $m/z$ 77 could then be produced by loss of methyl radical.

The nature of the butyl group is uncertain, although the hydrocarbon ion series lacks an ion at $m/z$ 43. This mitigates strongly against the presence of either an *n*-propyl or, particularly, an isopropyl group in the molecule and leaves us with only *t*-butyl chloride (2-methyl-2-chloropropane) and *sec*-butyl chloride (2-chlorobutane) as possible answers. Spectra of both of these compounds would be desirable to help distinguish between them, although consideration of the probable α-cleavage fragmentations of these compounds (see Chapter 5) makes *t*-butyl chloride the most reasonable choice [*t*-butyl chloride; 2-methyl-2-chloropropane]:

(4.8)

**Problems 4.3 through 4.7.** Identify the compounds whose spectra are given in Figures 4.18 through 4.22. It may not always be possible to narrow the possibilities down to a unique choice, but you should be able to support whatever choice or choices you make. (For Problem 4.6, it is important only to determine what *type* of compound would give rise to this spectrum. A full discussion will be given in Chapter 5.)

## REFERENCES

R.M. Smith, *Anal. Chem.*, **54**, 1399A (1982).

R.M. Smith in M.H. Ho, Ed., *Analytical Methods in Forensic Chemistry*, Ellis Horwood Ltd., Chichester, U.K., 1990, pp. 16–28.

R.M. Smith in J. Yinon, Ed., *Forensic Mass Spectrometry*, CRC Press, Boca Raton, FL, 1987, pp. 30–60.

R.M. Smith, *J. Forensic Sci.*, **28**, 318 (1983).

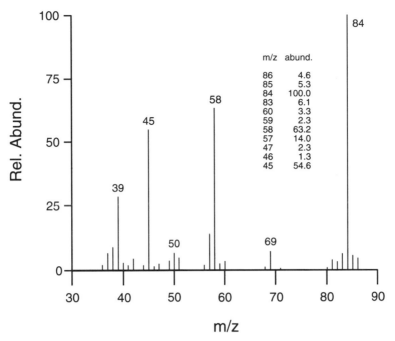

**Figure 4.18.** Mass spectrum for Problem 4.3.

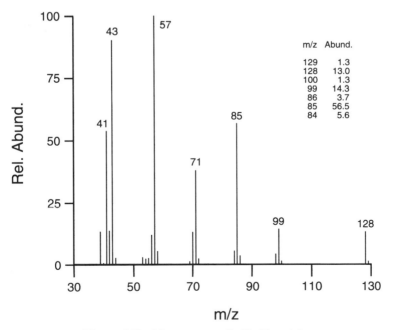

**Figure 4.19.** Mass spectrum for Problem 4.4.

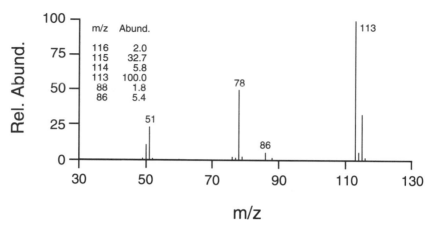

**Figure 4.20.** Mass spectrum for Problem 4.5.

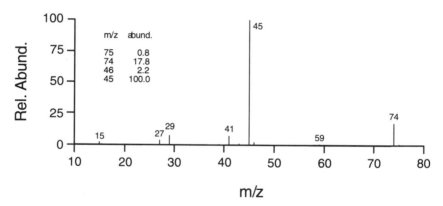

**Figure 4.21.** Mass spectrum for Problem 4.6.

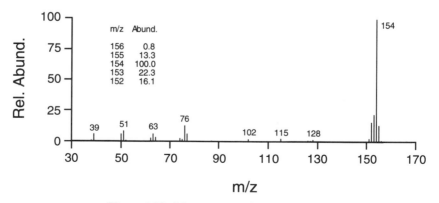

**Figure 4.22.** Mass spectrum for Problem 4.7.

# CHAPTER 5

# ALPHA-CLEAVAGE

## 5.1. INTRODUCTION

In the last chapter several fragmentations of the molecular ion were introduced—some involving cleavage of a single $\sigma$-bond, others involving more complex rearrangement of the $\pi$-electron system. In this chapter we will focus on one specific type of fragmentation in which one $\sigma$-bond is cleaved, and rearrangement of electron density occurs to help stabilize the positive charge. This fragmentation is called $\alpha$-cleavage because cleavage occurs at a $\sigma$-bond to the carbon that is $\alpha$ (i.e., adjacent) to a heteroatom or a group that can act like a heteroatom. This fragmentation has also been called $\beta$-scission or $\beta$-cleavage in the mass spectral literature, referring to the fact that the *group* that is lost is one atom removed from the heteroatom.

Specific conditions must be met if $\alpha$-cleavage is to occur. In the representation shown in Figure 5.1, the heteroatom or heteroatom-like group is denoted by X; the R groups are usually either alkyl groups or hydrogen. As we saw in Section 3.2, molecules containing heteroatoms undergo initial ionization at the heteroatom through loss of a nonbonding ($n$) electron. Fragmentation then proceeds by cleavage of one of the $\sigma$-bonds to the $\alpha$-carbon (arbitrarily shown here as the $C-R^1$ bond), with one of the $\sigma$-bonding electrons leaving with $R^1$. Following the logic discussed in Section 3.3, because the ions formed from loss of the various R groups have similar stabilities, the R group forming the most stable radical will be lost preferentially. The $\alpha$-carbon then rehybridizes to accommodate the remaining $\sigma$-bonding electron in a $p$-orbital, which overlaps with the unpaired, nonbonding electron on the heteroatom to form a $\pi$-bond between the carbon and the X group.

The X group must possess two properties in order to direct $\alpha$-cleavge. First, it must be more electronegative than carbon so that it withdraws $\sigma$ electron density away

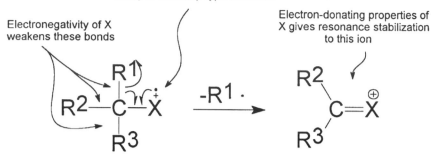

**Figure 5.1.** Generalized representation of α-cleavage fragmentation. X = NR$_2$, OR, double bond or aromatic ring; or X, R$^3$ = =O. R group forming most stable radical will be lost preferentially.

from the other bonds to the α carbon, thereby weakening them. Second, and just as important, X must have $n$- or π-electrons available to donate back to the α-carbon to help stabilize by resonance the positive charge that develops:

$$\oplus CR_2 - X \leftrightarrow R_2C = X\oplus$$

where the charge is shared unequally between the α-carbon and the X group. If X is an aromatic ring, a π-electron may be lost in initial ionization from the ring (Figure 5.2).

The ability of X to stabilize the positive charge at first may seem to contradict the fact that X must also be fairly electronegative. Groups that best support α-cleavage

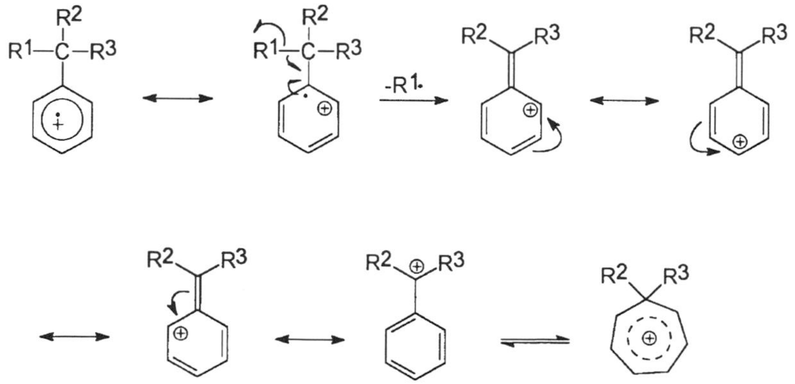

**Figure 5.2.** α-Cleavage with an aromatic ring as the X group, showing charge stabilization by the ring.

are those whose electronegativities are closest to that of carbon, since they most comfortably share the positive charge. Nitrogen is so well adapted to this role that molecules containing aliphatic nitrogen undergo $\alpha$-cleavage sometimes to the near exclusion of other fragmentation reactions. When nitrogen is contained in a complex saturated ring system, many of the most intense ions in the spectrum result from rearrangements in which the positive charge is stabilized by the nitrogen (e.g., see cocaine, Section 8.3).

Several common X groups meet the two criteria in Figure 5.1: the oxygen in alcohols and ethers, the nitrogen in amines, and any aromatic ring not loaded with electron-donating groups. The doubly bonded oxygen of a carbonyl group is also included (but not the entire carbonyl group!), although in this case the symbolism necessitates removal of one of the R groups from the carbonyl carbon (Figure 5.1). As with other X groups, the nonbonding electrons on the carbonyl oxygen stabilize the positive charge by forming an additional $\pi$-bond to the carbonyl carbon ($R^2-C\equiv O\oplus$). Although these groups are the most common directors of $\alpha$-cleavage, halogens (with the exception of fluorine), the sulfur in mercaptans (RSH) and sulfides (RSR), and even isolated double bonds (vinyl groups; $R_2C=CR-$) also can cause this fragmentation to occur at an adjacent carbon atom.

## 5.2. BENZYLIC CLEAVAGE

In Figure 5.2, an aromatic compound having carbon next to the ring undergoes $\alpha$-cleavage so that, with the exception of its bond to the ring, the bonds to that carbon (the benzylic carbon) are broken. The spectra in Figure 5.3 illustrate this type of fragmentation, which we will call benzylic cleavage, for three isomeric alkylated benzenes. In each case the largest possible alkyl radical is lost (isopropyl, ethyl, and methyl, respectively) to give the base peaks at $m/z$ 91, 105, and 119. The only other losses possible by benzylic cleavage in these molecules are hydrogen radicals. In isobutylbenzene (Figure 5.3a), the stability of hydrogen radical cannot compete with that of isopropyl radical, so that the $(M-1)^+$ ion is not observed. On the other hand, 3,5-dimethylethylbenzene (Figure 5.3c) has eight benzylic hydrogen atoms (from three benzylic carbons), and by sheer statistical force the $(M-1)^+$ ion becomes detectable, even though methyl loss is still preferred by a factor of about 20. Benzylic cleavage is *the* most important fragmentation these molecules undergo, reflecting the much smaller activation energies required by $\alpha$-cleavage when compared to other fragmentations of aromatic rings.

Benzylic cleavage may be significant even in more complex molecules, especially if the resulting ion is stabilized by conjugation with several aromatic rings or with heteroatom functional groups on the ring. For example, papaverine, a potent nonnarcotic alkaloid found in opium, contains a carbon atom that is "doubly benzylic"— that is, attached to two different aromatic rings. In the mass spectrum of this compound (Figure 5.4), loss of hydrogen radical from this carbon produces the base peak at $m/z$ 338 because of the large number of resonance structures that stabilize the resulting ion. In this case, initial ionization occurs preferably at the heteroatoms [Eq. (5.1)], rather than in the aromatic ring itself (compare Figure 5.2).

(5.1)

m/z 339          m/z 338

m/z 308

In addition to the oxygen in the *para* position of the dimethylcatechol ring, stabilization of the positive charge is possible at the *ortho, meta,* and *para* carbons of this ring, at five carbons and one of the oxygens in the dimethoxyisoquinoline ring system, and on the benzylic carbon itself. The oxygens on these rings do not provide the sole stabilization for this ion, however, since the spectrum of diphenylmethane ($\phi_2CH_2$) exhibits an $(M - 1)^+$ ion at *m/z* 167 having an intensity of 90% relative to that of the molecular ion (*m/z* 168; 100%).

The ion at *m/z* 308 in Figure 5.4 occurs by characteristic loss of formaldehyde from one of the aromatic methoxy groups [Eq. (5.1); compare Eq. (4.5)]. The other intense peak in the spectrum (*m/z* 324) is produced by methyl radical loss from one of the methyl ether groups (Section 5.4.1).

If the aromatic ring is crowded with substituents, relief of steric strain in the molecular ion may override the tendency toward benzylic cleavage. 1,2,3,5-Tetramethylbenzene, for example, loses methyl to produce the base peak at *m/z* 119 (Figure 5.5*a*), presumably by loss of the methyl group in the 2-position. In an extreme situation, the four adjacent methyl groups of 1,2,3,4-tetramethylbenzene produce so much strain that the molecule virtually falls apart when ionized (Figure 5.5*b*). Other than the aromatic low mass ion series, this mass spectrum tells very little about the original structure of the molecule.

**Problem 5.1.** The two compounds for which spectra are shown in Figure 5.6 have the same molecular weight. Relative abundance data for the molecular ion and the base peak of each spectrum follow:

| a | | b | |
|---|---|---|---|
| m/z | Relative Abundance | m/z | Relative Abundance |
| 121 | 2.3 | 121 | 2.4 |
| 120 | 22.8 | 120 | 23.6 |
| 92 | 10.5 | 106 | 9.2 |
| 91 | 100.0 | 105 | 100.0 |

Identify the compounds that gave rise to these spectra. If necessary, refer back to Section 2.3 to organize your approach to these problems.

**Problem 5.2.** The chromatogram in Figure 5.7 was obtained from methanolic dilution of the contents of a spray canister labeled "CS Tear Gas" (*ortho*-chlorobenzal-

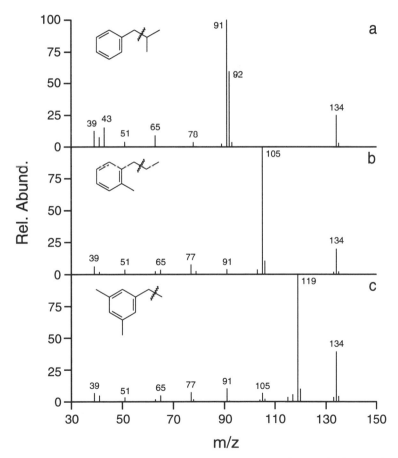

**Figure 5.3.** Mass spectra of three $C_4$-benzenes. In each case the largest alkyl group is lost by benzylic cleavage from a carbon next to the ring.

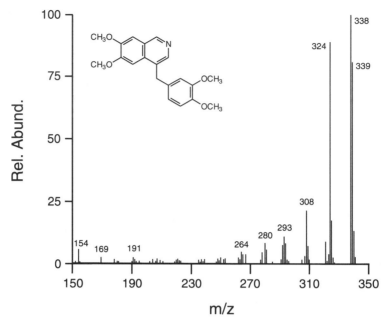

**Figure 5.4.** Mass spectrum of papaverine. The (M − 1)⁺ ion is the base peak due to loss of the "doubly benzylic" hydrogen.

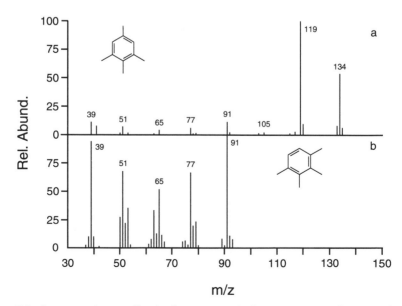

**Figure 5.5.** Severe steric crowding in these tetramethylbenzenes causes fragmentation that overrides benzylic cleavage.

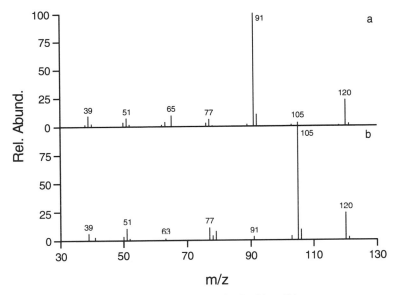

**Figure 5.6.** Mass spectra for Problem 5.1.

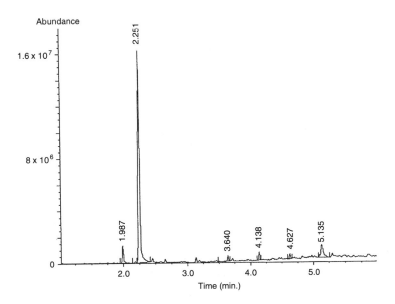

**Figure 5.7.** Total ion chromatogram of an extract of the contents of a "CS" tear gas canister (Problem 5.2).

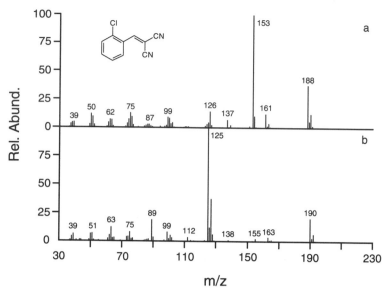

**Figure 5.8.** Mass spectra of *o*-chlorobenzalmalononitrile ("CS" tear gas; top) and an unknown impurity (Problem 5.2). (*a*) Scan 216 (2.254 min.); (*b*) scan 170 (1.989 min.).

malononitrile). The small peaks at high retention times come from the mineral oil carrier in the can. Mass spectra from the peaks at 1.987 and 2.251 min are shown in Figure 5.8. The largest of these peaks was identified as CS by library search, but the smaller peak gave no computer matches. Identify the major losses in the CS mass spectrum, and convince yourself that the library search results are correct. Then determine the structure of the other compound.

## 5.3. CLEAVAGE NEXT TO ALIPHATIC NITROGEN

Two general guidelines are worth remembering about compounds containing aliphatic nitrogen: (1) these compounds rarely break the carbon–nitrogen bond, and even when such fragmentation occurs, it usually accounts for only a small fraction of the total fragmentation and (2) the base peak in the spectra of most of these compounds arises either directly via α-cleavage or, when the nitrogen is contained in a ring system, by initial α-cleavage and subsequent rearrangement that keeps the positive charge on the nitrogen.

The mass spectrum of amphetamine, a powerful central nervous system stimulant, illustrates the importance of α-cleavage in these compounds. The outstanding feature in this spectrum (Figure 5.9) is the base peak (*m/z* 44), which is over 10 times more intense than any other ion in the spectrum. The ions at *m/z* 50 to 52, 63 to 64, 77 to 78, and 89 to 90 constitute the aromatic low mass ions series, with a benzylic series at *m/z* 91 and 65 (ions below *m/z* 40 are not shown). Of the remaining ions, none are greater than about 3% relative abundance. The (M − 1)$^+$ ion at *m/z* 134 is larger than

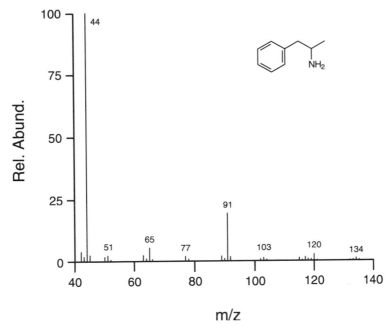

**Figure 5.9.** Mass spectrum of amphetamine. This type of fragmentation pattern, with one extremely intense low mass ion and few other ions of significant size, is characteristic of many aliphatic amines.

the molecular ion at $m/z$ 135 (remember the odd-nitrogen rule—amphetamine contains an odd number of nitrogens and thus *must* have an odd molecular weight).

Ignoring for the moment the small peaks in the regions between $m/z$ 102 to 104 and 115 to 119, which are alkylbenzene ions arising from loss of the $NH_2$ group, let us concentrate specifically on the ions at $m/z$ 134, 120, 91, and 44, all of which are produced by α-cleavage. The three ions with the positive charge stabilized on nitrogen ($m/z$ 44, 120, and 134) should have fairly similar stabilities since there are no inherently stabilizing or destabilizing groups attached to the nitrogen, and the aromatic ring cannot offer assistance by conjugation. The only factor remaining to explain the differences in the ion abundances is the relative stabilities of the radicals formed in the same reactions [Eq. (5.2); see Sections 3.3 and 4.2.1].

Although $m/z$ 91 (benzyl ion) can arise by the sort of mechanism depicted in Figure 5.2 for the alkylbenzenes, this does not take into account the fact that initial ionization in amphetamine is more likely to occur at nitrogen. Thus, while formation of the $m/z$ 44, 120, and 134 ions in amphetamine and of the benzyl ion in Figure 5.2 are all processes in which the charge is retained at the initial site of ionization (charge retention), benzyl ion formation in amphetamine results from moving the positive charge from the nitrogen atom to the aromatic ring [charge migration; Eq. (5.3)]. Because additional energy is required to move the charge from one site to another during charge migration, this ion is less intense despite its stability (Section 3.3).

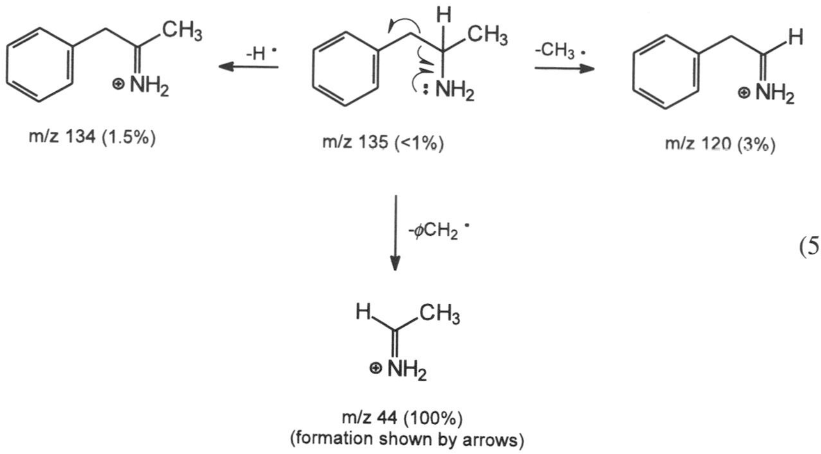

m/z 134 (1.5%)          m/z 135 (<1%)          m/z 120 (3%)

(5.2)

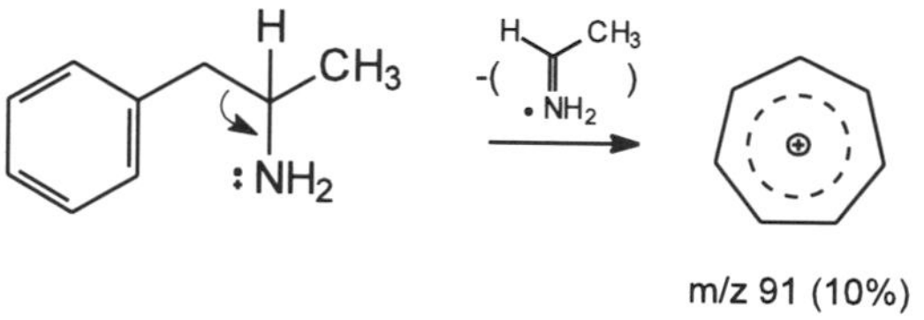

m/z 44 (100%)
(formation shown by arrows)

(5.3)

m/z 91 (10%)

Except for the ions at $m/z$ 44 and 91 in Figure 5.9, the other ions formed by α-cleavage are so small as to be easily overlooked. However, the intensities of these small ions do in fact reflect structural differences in these molecules. The spectra of phentermine and methamphetamine, two isomeric stimulants with obvious structural similarities to amphetamine, illustrate this point well (Figure 5.10). In this figure the base peak in each spectrum ($m/z$ 58) is actually five times larger than shown, which accounts for the (X5) located next to the 58 in each spectrum and for the fact that the $y$ axis in each spectrum only goes up to 20% relative abundance. Using this format for presentation accentuates the differences in abundance between the less intense ions in the spectra.

Although there are several differences between these spectra at low mass, at high mass only the intensities of the ions at $m/z$ 134 (relative to the ion at $m/z$ 91) and the presence of a tiny ion at $m/z$ 148 in the methamphetamine spectrum distinguish the two. These differences are reproducible, and as such are indicative of the structural differences between the two molecules.

During α-cleavage, phentermine can lose either of the *geminal* methyl groups to give the ion at $m/z$ 134 [Eq. (5.4)], benzyl radical to give the base peak at $m/z$ 58, or benzyl ion ($m/z$ 91) by charge migration:

(5.4)

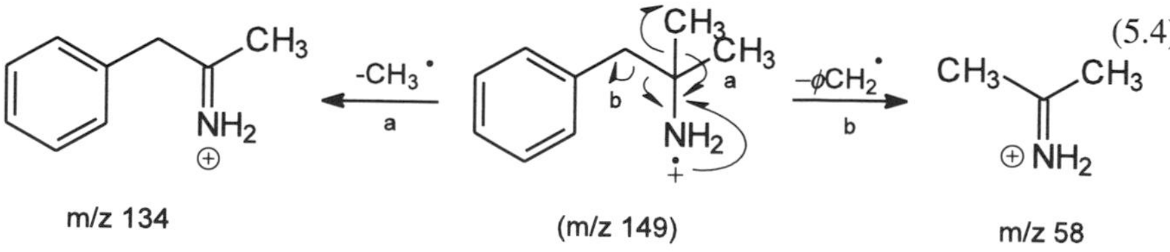

m/z 134          (m/z 149)          m/z 58

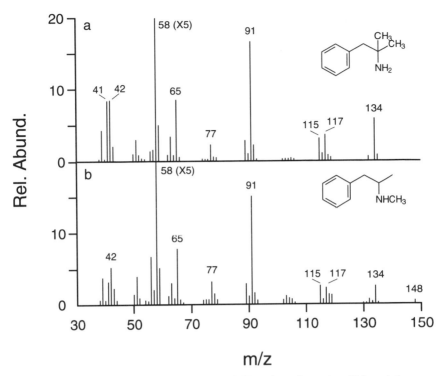

**Figure 5.10.** Mass spectra of (*a*) phentermine and (*b*) methamphetamine. Although these compounds have similar structures and extremely similar mass spectra, subtle differences in the patterns of α-cleavage distinguish them. In the actual mass spectra, the *m/z* 58 ions are five times larger than shown.

Although phentermine theoretically could lose hydrogens from the benzylic carbon by α-cleavage, this loss is only observed when a substantial number of these hydrogens are present, and then only if other fragmentations are not more highly favored (Section 5.2). In addition, loss of the benzylic hydrogens in phentermine can occur only by initial ionization in the aromatic ring, which is much less likely than ionization at the nitrogen atom.

Methamphetamine, on the other hand, has two carbons next to the nitrogen—one having three hydrogens (the *N*-methyl group), the other with a hydrogen, a methyl group, and a benzyl group. Loss of any of the four hydrogens gives rise to the ion at *m/z* 148, whereas loss of the aliphatic methyl produces the ion at *m/z* 134, loss of benzyl radical the ion at *m/z* 58, and charge migration α-cleavage leads to the benzyl ion (*m/z* 91) [Eq. (5.5)]. The presence of the tiny *m/z* 148 ion in the methamphetamine spectrum, then, is not accidental; it is predictable on the basis of the molecular structure. In comparing the two spectra in Figure 5.10, even the difference in the relative sizes of the *m/z* 134 ion seems to reflect the fact that phentermine has two methyl groups that can be lost by α-cleavage as opposed to only one for methamphetamine, especially since the intensities of *m/z* 91 are similar in the two spectra. Loss of the methylamino group by cleavage of the carbon–nitrogen bond leads to the alkylbenzene ions

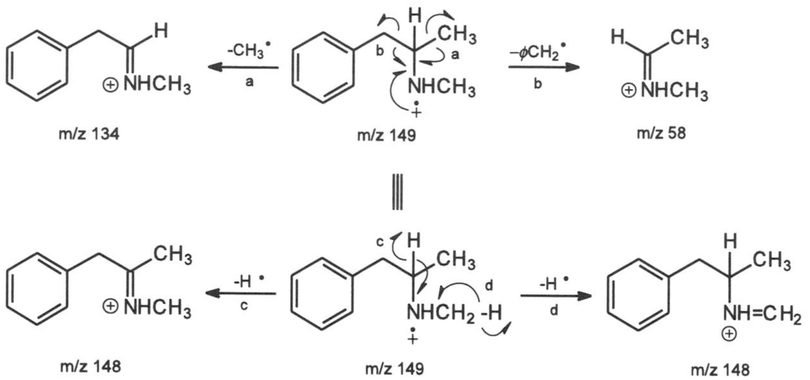

(5.5)

m/z 134          m/z 149          m/z 58

m/z 148          m/z 149          m/z 148

at $m/z$ 115 to 119 ($m/z$ 117 is $\phi-CH=CH-CH_2\oplus \leftrightarrow \phi-CH\oplus-CH=CH_2$), but this fragmentation does not compete well with the major modes of α-cleavage.

At this point the origin of the low mass ion series for aliphatic amines (Table 4.2) should be clear. The spectra of amphetamine, methamphetamine, and phentermine contain intense ions at $m/z$ 44 and 58 that dwarf all other ions in the spectra. Compounds having more, or larger, aliphatic groups attached to the nitrogen atom still fragment so as to lose the largest aliphatic group and produce an intense ion (usually the base peak) in which the positive charge is located on the nitrogen. Thus *N,N*-dimethylamphetamine [$\phi CH_2CH(CH_3)N(CH_3)_2$], which contains one more methyl group on the nitrogen than does methamphetamine, produces a base peak at $m/z$ 72:

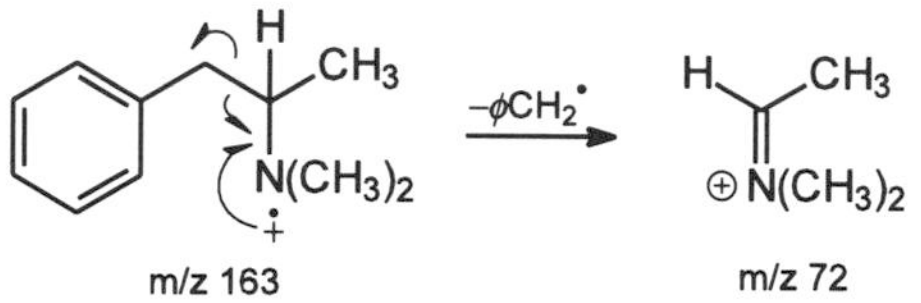

(5.6)

m/z 163          m/z 72

And *N,N*-dimethylphentermine [$\phi CH_2C(CH_3)_2N(CH_3)_2$], containing two more methyl groups than phentermine, gives a base peak at $m/z$ 86:

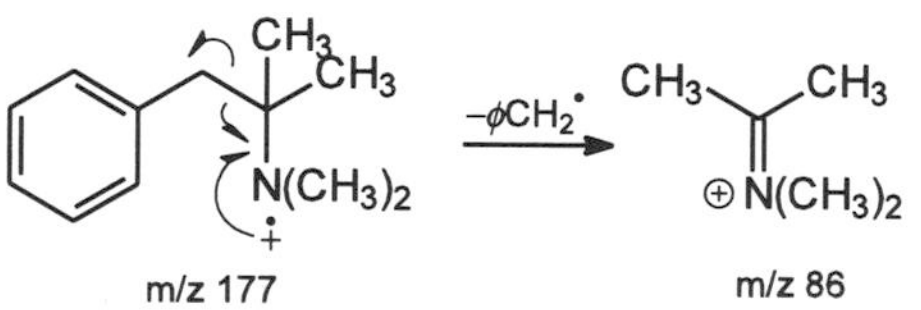

(5.7)

m/z 177          m/z 86

The presence of an intense ion at any of these masses ($m/z$ 44, 58, 72, 86, . . . ) is an almost immediate clue that somewhere in the molecule is an aliphatic group containing nitrogen. This is true even for fairly complex molecules such as lidocaine, a local anesthetic, and amitriptyline, an antidepressant drug (Figure 5.11).

***Problem 5.3.*** The five compounds that produced the spectra shown in Figure 5.12 are all isomers of methamphetamine. (i) The ion at $m/z$ 91 is fairly prominent in all of these spectra. What does this indicate about the possibility of substitution on the

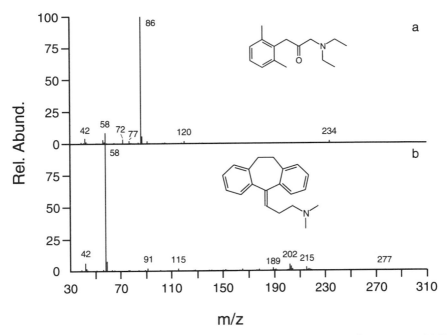

**Figure 5.11.** Mass spectra of two aliphatic amines having complex aromatic structures: (*a*) lidocaine and (*b*) amitryptilene. α-Cleavage still dominates the fragmentation in these molecules.

aromatic rings of these compounds? What would happen to this ion if any of these compounds had additional alkyl groups located on the benzylic carbon? (ii) The base peak in each spectrum is *m/z* 58. What does this tell us about the distribution of alkyl groups in these molecules? Draw out all of the possible isomeric structures that meet the criteria in parts (i) and (ii). (iii) Determine which spectrum is that of methamphetamine by comparison with Figure 5.10. Predict the products of α-cleavage for each of these compounds. Finally, match the predicted fragmentation patterns to those observed in the spectra in Figure 5.12.

***Problem 5.4.*** The compound whose structure is given below was reported as a by-product of methamphetamine synthesis, but no electron ionization mass spectrum of the compound was included in the article. Predict what this spectrum might look like by making a list of all the possible ions and radicals that would result from the various α-cleavages. Next, arrange the ions (do not worry about the radicals right now) into groups of similar stability, and rank them from least to most stable on the basis of such factors as conjugation. Finally, for ions in the most stable group, rank the cor-

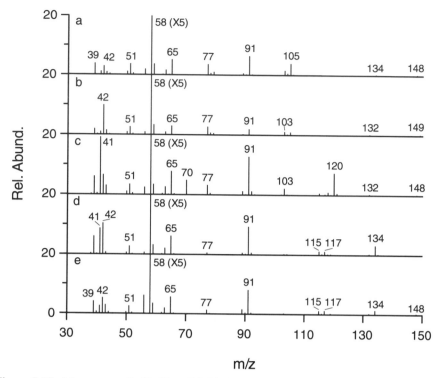

**Figure 5.12.** Mass spectra for Problem 5.3. The intensities of the m/z 58 ions are actually five times greater than shown.

responding radical products from least to most stable. If the base peak in the spectrum is determined by a fragmentation that forms the most stable pair of products, what will be the base peak in this spectrum?

The X group in Figure 5.1 must be able to donate π-type electrons back to the carbon to help stabilize the positive charge, an ability that is severely compromised when strong electron-withdrawing groups are present on the nitrogen. Compare, for example, the spectra of N-pentafluoropropionylnortriptylene and the parent drug nortriptylene (Figure 5.13), the latter of which shows a base peak at m/z 44 from α-cleavage and little additional fragmentation:

$$\qquad\qquad\qquad\qquad\qquad\qquad\qquad\qquad\qquad (5.8)$$

m/z 232 (100%;
R=COCF$_2$CF$_3$ only)

m/z 44 (100%; R=H)
m/z 190 (12%;
R=COCF$_2$CF$_3$)

α-Cleavage contributes only a minor peak to the spectrum of the pentafluoropropionyl

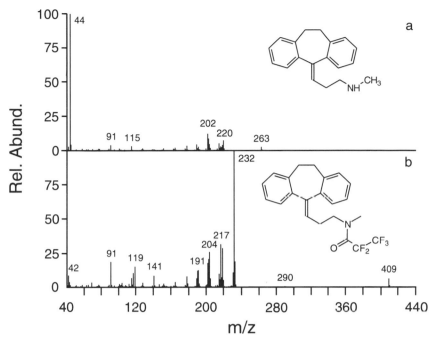

**Figure 5.13.** Mass spectra of (*a*) nortriptylene and (*b*) its *N*-pentafluoropropionyl derivative. A strongly electronegative group greatly reduces nitrogen's ability to support the positive charge during α-cleavage.

derivative (*m/z* 190; 12%), with the base peak in the spectrum coming from a McLafferty-type γ-hydrogen rearrangement [Eq. (5.8); see Section 6.2.1]. The incipient transition state for α-cleavage is so destabilized by the lack of available π-electrons, here conjugated with the highly electronegative pentafluoropropionyl group, that other fragmentations become more favorable. Even transformation of an amine to a formyl amide substantially reduces the relative size of α-cleavage ions [Figure 5.14; Eq. (5.9)]:

$$(5.9)$$

Again, the base peak in the spectrum is due to a McLafferty-type γ-hydrogen rearrangement.

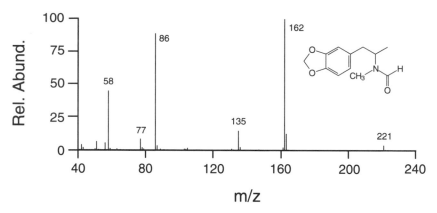

**Figure 5.14.** Mass spectrum of *N*-formyl-3,4-methylenedioxymethamphetamine. Even an unsubstituted carbonyl on nitrogen allows other fragmentations to compete favorably with α-cleavage (compare Figure 8.2*b*).

## 5.4. CLEAVAGES OF ALIPHATIC OXYGENATED COMPOUNDS

### 5.4.1. Alpha-Cleavage

Like nitrogen, the oxygens of aliphatic alcohols and ethers are strong directors of α-cleavage. However, because oxygen is more electronegative than nitrogen, it is less able to support a positive charge and, as a result, other fragmentations are sometimes more important than α-cleavage in these compounds.

Acetaldehyde dimethylacetal does not produce a molecular ion (Figure 5.15; the apparent first loss of 14 daltons from *m/z* 89 to 75 is inconsistent with *m/z* 89 being the molecular ion). In fact, aliphatic oxygenated compounds are notorious for producing very weak molecular ions since the electronegativity of oxygen weakens some bonds enough that activation energies for their breakage are extremely low.

Experience with α-cleavage now allows us to predict probable losses from this molecule. There are actually three carbons next to the two oxygen atoms in this molecule, two of which contain only hydrogens:

$$
\underset{\substack{\text{m/z 89}}}{\overset{\substack{CH_3 \\ |}}{CH_3O-\overset{\oplus}{C}=OCH_3}}
\quad \xleftarrow{-H^\bullet} \quad
\underset{\substack{\text{m/z 90 (not observed)}}}{\overset{\substack{CH_3 \\ |}}{CH_3O-\underset{\underset{H}{|}}{C}-OCH_3}}
\quad \xrightarrow{-CH_3^\bullet} \quad
\underset{\substack{\text{m/z 75}}}{\overset{\substack{CH_3 \\ |}}{CH_3O-\underset{\underset{H}{|}}{C}=\overset{\oplus}{O}CH_3}}
\tag{5.10}
$$

$$\downarrow {-CH_3O^\bullet}$$

$$
\underset{\substack{\text{m/z 59 (formation} \\ \text{shown by arrows)}}}{\overset{\substack{CH_3 \\ |}}{\underset{\underset{H}{|}}{C}=\overset{\oplus}{O}CH_3}}
$$

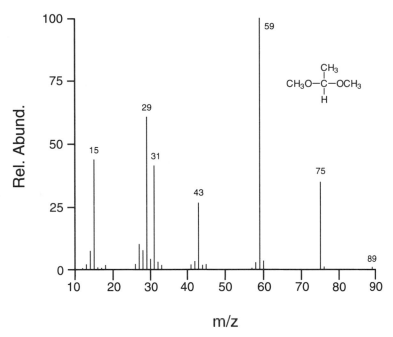

**Figure 5.15.** Mass spectrum of 1,1-dimethoxyethane. Although oxygen also directs α-cleavage, the effect is not as pronounced as with nitrogen.

However, the presence of seven hydrogens that can be lost by α-cleavage increases the likelihood of seeing the $(M - 1)^+$ ion at $m/z$ 89. Of the two groups remaining on the central carbon, it is not clear beforehand which group will likely be lost. On the basis of size alone, loss of the methoxy radical might seem more reasonable, yet the oxygen provides no obvious additional stability to methoxy radical that would account for the threefold difference in abundance observed in Figure 5.15. A more likely explanation is that the ion resulting from methyl loss still contains two oxygens, and while one oxygen stabilizes the positive charge, the other drains electron density away from the same area because of its electronegativity. This overrides any stability gained from having the two oxygens share the positive charge by resonance and *destabilizes* formation of the $m/z$ 75 ion relative to the $m/z$ 59 ion resulting from methoxy radical loss, in which the remaining methyl group, being slightly electropositive, actually helps stabilize the ion product.

It is instructive at this point to return to Problem 4.6 (Figure 4.21). In the context of the previous chapter, it was possible only to classify this spectrum as that of either an aliphatic alcohol or ether because the intense $m/z$ 45 ion fit into the low mass ion series for this group of compounds. Many beginning students guess that the answer to this problem is diethylether ($CH_3CH_2OCH_2CH_3$) because of the apparent loss of ethyl from the molecular ion at $m/z$ 74 to give the base peak at $m/z$ 45. It should now be apparent, however, that diethylether cannot be the correct solution to this problem since both carbon atoms next to the oxygen contain only hydrogen and methyl. Assuming that α-cleavage is the primary mode of fragmentation, diethylether should give rise to a base peak at $m/z$ 59 from loss of methyl radical. In fact, we expect to see only a very small peak in the spectrum of diethylether due to loss of an ethyl group.

Of the seven isomeric structures that can be written for aliphatic alcohols and ethers having the empirical formula $C_4H_{10}O$, only two can lose ethyl radical by α-cleavage: *sec*-butanol $[CH_3CH_2CH(OH)CH_3]$ and methyl *n*-propyl ether $(CH_3CH_2CH_2OCH_3)$. Thus the compound responsible for the spectrum in Problem 4.6 must be one or the other of these two compounds. Although the low intensity of the $m/z$ 59 ion (from loss of methyl) might suggest at first that methyl propyl ether is the more likely candidate, only a comparison of the spectra of these two compounds can answer this question unambiguously (see Problem 5.6 at the end of this chapter).

Combining an aliphatic ether and an aromatic system increases the likelihood of α-cleavage. Cannabinol, the naturally occurring oxidation product of tetrahydro-cannabinols in marijuana, readily loses one of the *geminal* methyl groups that are situated not only on a benzylic carbon but also on a carbon next to oxygen. The resulting ion has a fully aromatized structure:

(5.11)

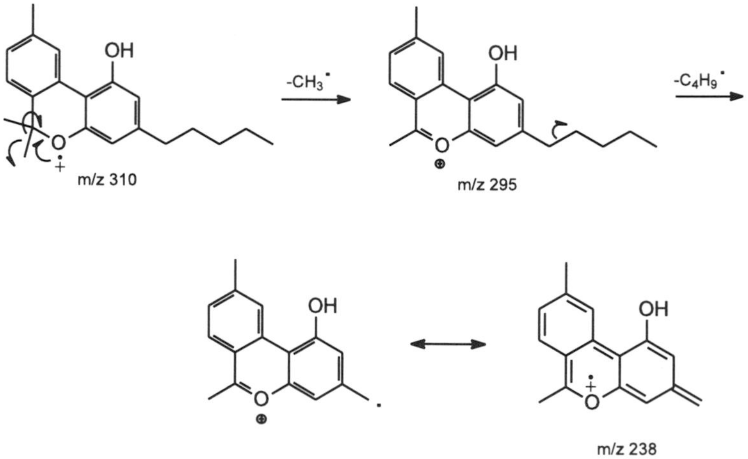

and gives rise to the base peak in the spectrum at $m/z$ 295 (Figure 5.16). Subsequent benzylic-type cleavage leads to the only other ion of significance in the spectrum at $m/z$ 238.

Loss of methyl from the molecular ion of papaverine to give the intense ion at $m/z$ 324 (Figure 5.4) can be viewed as "α-cleavage by long distance." In this case, not only is the lost methyl group not located on the carbon next to the initially ionized oxygen, but the fragmentation also occurs by breaking a carbon–heteroatom bond!

(5.12)

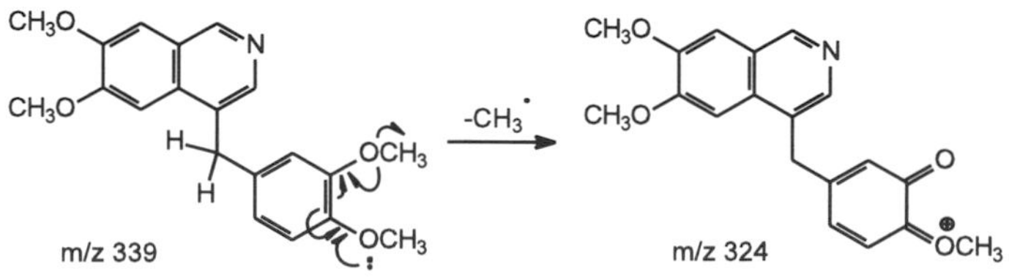

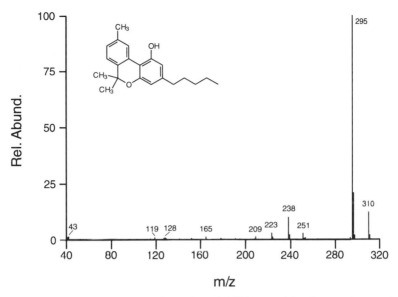

**Figure 5.16.** Mass spectrum of cannabinol. The $m/z$ 295 ion resulting from α-cleavage has extensive aromatic stabilization.

However, the resulting rearrangement of the electronic structure produces an ion that remains stabilized by the extended aromatic system while retaining the positive charge on the initially ionized oxygen.

## 5.4.2. Water Elimination in Aliphatic Alcohols

Primary alcohols with a chain length of four carbons or longer are prone to loss of water by formation of a cyclic intermediate. The spectrum of $n$-pentanol in Figure 5.17 exhibits a base peak at $m/z$ 42, with other important fragment ions at $m/z$ 55 and 70—none of which are expected if α-cleavage is the primary mode of fragmentation. In comparison, the spectrum of $n$-pentamine, the corresponding nitrogen-containing compound,

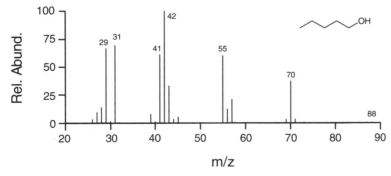

**Figure 5.17.** Mass spectrum of $n$-pentanol is dominated by fragmentations related to the loss of water.

shows a base peak at $m/z$ 30 from $\alpha$-cleavage, a molecular ion at $m/z$ 87 (about 7% relative intensity), and no other fragment ions having intensities greater than 1 to 2%.

How do we account for this difference? First, it is important to understand that loss of water by these compounds during mass spectral fragmentation is very different from that observed under acid-catalyzed conditions in solution. In the latter case, the hydroxyl group is protonated, water is lost to form a carbocation, and hydrogen is removed from an adjacent carbon to form a double bond:

$$(5.13)$$

Secondary and tertiary alcohols are more likely to lose water by this reaction than are primary alcohols because the carbocations formed are more stable. Detailed studies of the mass spectra of aliphatic alcohols show, however, that only primary alcohols undergo this fragmentation! Furthermore, the hydrogen lost in addition to the hydroxyl group comes not from an adjacent carbon but rather from a position several carbons removed, typically through formation of a six-membered cyclic intermediate:

$$(5.14)$$

This rearrangement is aided by the electronegativity of the ionized oxygen in the molecular ion and by the strength of the oxygen–hydrogen bond that is formed. In aliphatic amines, the less electronegative nitrogen is not as attractive to hydrogen and is better able to stabilize the positive charge in the $\alpha$-cleavage ion, reducing the importance of ammonia loss. In secondary and tertiary alcohols, the $R_2C=OH\oplus$ ions formed by $\alpha$-cleavage are stabilized by the additional alkyl substituents, making $\alpha$-cleavage the more attractive mode of fragmentation.

Once formed, the $m/z$ 70 ion from $n$-pentanol reacts like other primary carbocations in straight-chain environments (see Section 4.2.1), losing ethylene to give the more intense ion at $m/z$ 42 [Eq. (5.14)].

### 5.4.3. Cleavage at Carbonyl Groups

The oxygen of a carbonyl group also directs $\alpha$-cleavage at the adjacent carbon—in this case, the carbonyl carbon. Since only two groups are bonded to this carbon (the third bond being the carbon–oxygen $\pi$-bond), cleavage occurs on either side of the carbonyl carbon. For aliphatic ketones, these losses are usually straightforward. Both losses are observed, but that of the larger alkyl radical leads to the most intense fragment ion, in keeping with previous discussions. The spectrum of 3-pentanone (Figure 4.4a) illustrates the situation when both alkyl groups are the same; the single propionyl ion at $m/z$ 57 resulting from $\alpha$-cleavage ($CH_3CH_2C\equiv O\oplus$) is the most intense ion in the spectrum (see also Example 4.1).

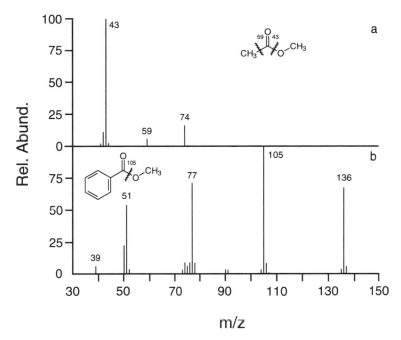

**Figure 5.18.** Mass spectra of an aliphatic and an aromatic methyl ester: (*a*) methyl acetate and (*b*) methyl benzoate. Note the intense benzoyl ion series in (*b*).

Other types of carbonyl compounds undergo this fragmentation as well, with aliphatic and aromatic compounds sometimes exhibiting different behaviors. In the spectra of methyl acetate and methyl benzoate (Figure 5.18), the aliphatic compound shows loss of both groups attached to the carbonyl carbon—methyl radical to give *m/z* 59 and methoxy radical to give *m/z* 43. The preferred loss of the methoxy group follows the same logic as discussed previously with the spectrum of acetaldehyde dimethylacetal (Section 5.4.1). With methyl benzoate, there is no discernible loss of phenyl radical (*m/z* 59). Rather, methoxy radical is readily lost to produce the benzoyl ion ($\phi C \equiv O \oplus$), which fragments to give the very prominent benzoyl ion series at *m/z* 105, 77, and 51. Ions from other fragmentations of both of these compounds are of relatively low intensity.

Amides also undergo cleavage on either side of the carbonyl group. Indeed, the loss of 16 daltons from the molecular ion by primary amides is one of the few examples of losses of this mass (Table 4.1). Both aliphatic and aromatic primary amides exhibit this behavior, as shown in the spectra of *n*-butyramide and benzamide (Figure 5.19). The spectrum of *n*-butyramide also shows an ion at *m/z* 44 from loss of propyl radical. Although the argument can be made that propyl is larger and thus should be lost, comparison of radical size is not always applicable when comparing radical sites on different elements. A more likely explanation is that the ion formed after propyl radical loss obtains additional resonance stabilization of the positive charge on the amide nitrogen:

$$NH_2 - C \equiv O \oplus \leftrightarrow \oplus NH_2 = C = O$$

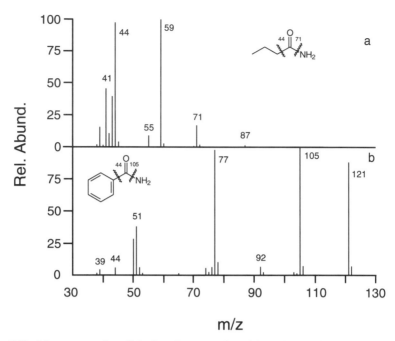

**Figure 5.19.** Mass spectra of an aliphatic and an aromatic amide: (*a*) butyramide and (*b*) benzamide.

The argument works in this case, where it did not with the loss of the methyl group in acetaldehyde dimethylacetal (Section 5.4.1), because the lowered electronegativity of nitrogen does not so severely destabilize the resonance form having the positive charge located on the carbonyl oxygen. In fact, nitrogen supports the positive charge better than oxygen does.

The other intense ion in the *n*-butyramide spectrum (*m/z* 59) arises from the McLafferty rearrangement, which is common in aliphatic carbonyl compounds (Section 6.2.1). Because of the structural requirements of the McLafferty rearrangement, a similar fragmentation was not observed in the spectrum of methyl acetate above. The spectrum of benzamide, like that of methyl benzoate, is dominated by the benzoyl ion series. The small, but noticeable, ion at *m/z* 44 corresponds to loss of phenyl radical. Although nitrogen helps support the positive charge in the $NH_2$-C$\equiv$O$\oplus$ ion, the aromatic ring, which has an even lower electronegativity and a greater capacity to stabilize by resonance, does an even better job.

Aliphatic and aromatic aldehydes have very different mass spectra (Figure 5.20). On the one hand, loss of the carbonyl hydrogen in *n*-pentanal cannot compete with loss of butyl radical to give *m/z* 29. Even the butyl ion at *m/z* 57, formed by charge migration, competes favorably for the positive charge. The base peak in this spectrum (*m/z* 44) is a McLafferty rearrangement ion.

In contrast, the benzoyl ion series, resulting from the loss of hydrogen radical from the molecular ion, dominates the spectrum of benzaldehyde. Loss of phenyl radical occurs only grudgingly in comparison (*m/z* 29 is less than 5% relative intensity), reflecting the relative stabilities of the benzoyl and the HC$\equiv$O$\oplus$ ions rather than the relative stabilities of the phenyl and hydrogen radicals.

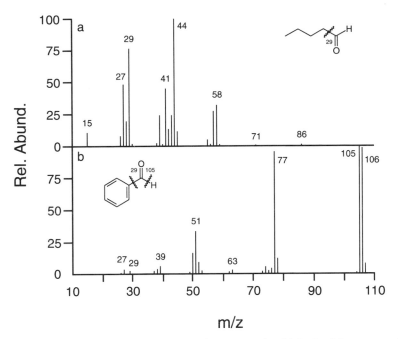

**Figure 5.20.** Mass spectra of an aliphatic and an aromatic aldehyde: (*a*) *n*-pentanal and (*b*) benzaldehyde.

## 5.5. SECONDARY ELIMINATION FROM INITIAL ALPHA-CLEAVAGE IONS

α-Cleavage alone does not explain the mass spectrum of ethyl isopropyl ether (Figure 5.21) since loss of the three methyl groups attached to the α carbons should lead to a base peak at *m/z* 73. Instead the spectrum implies, on the basis of what we have discussed in previous sections, that this molecule loses a propyl or isopropyl group by α-cleavage.

Analogous to the fragmentation of primary aliphatic ions (Section 4.2.1), certain ions formed by α-cleavage also fragment to give ions of similar stability by elimination of an olefin. This fragmentation occurs via a cyclic four-membered intermediate; therefore, certain structural features must be met: *if an ion formed by a-cleavage contains, across the heteroatom from the double bond, an alkyl group of two carbons or larger that has an available hydrogen on the second carbon from the heteroatom, this alkyl group will be lost as an olefin to produce an important ion from secondary cleavage.* This elimination is more prominent in ethers and alcohols (where it leads to ions that are more intense than the original α-cleavage ions) than in amines (in which the secondary ions are usually 50 to 70% as intense as the α-cleavage ions), probably for the reasons discussed in the previous section.

For ethyl isopropyl ether, the primary α-cleavage is indeed loss of methyl to give *m/z* 73. However, this ion has an ethyl group located across the heteroatom from the double bond that meets the structural requirement for secondary elimination to occur:

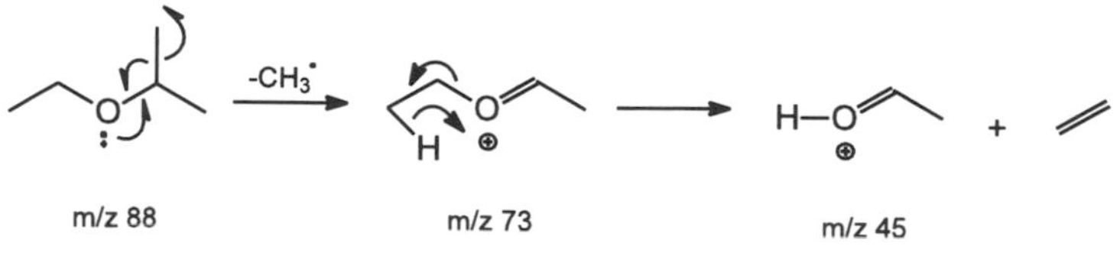

(5.15)

m/z 88                    m/z 73                    m/z 45

Actually, three other combinations for α-cleavage and subsequent secondary elimination are possible in this molecule:

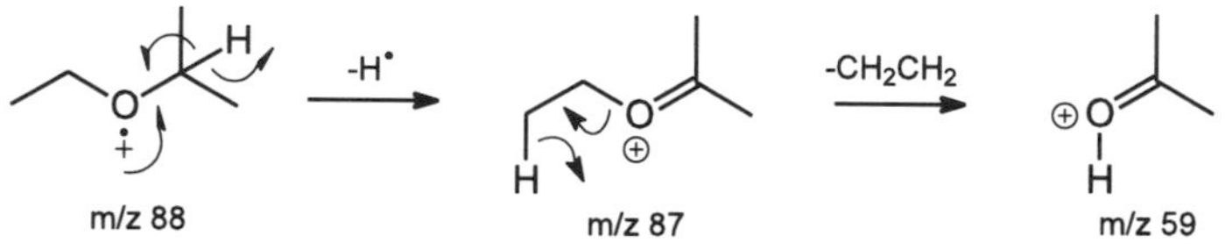

(5.16)

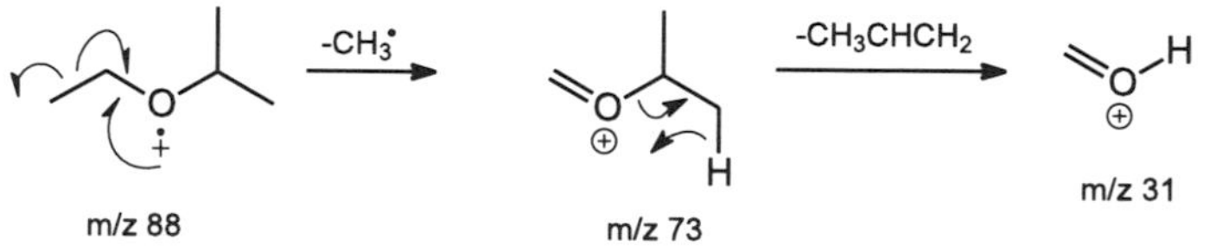

m/z 88              m/z 87              m/z 59

(5.17)

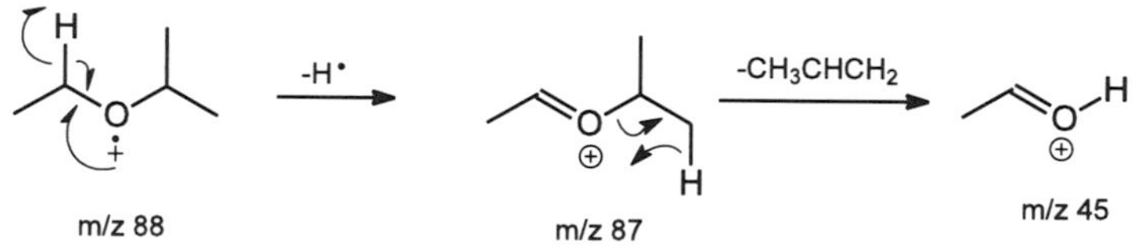

m/z 88              m/z 73              m/z 31

(5.18)

m/z 88              m/z 87              m/z 45

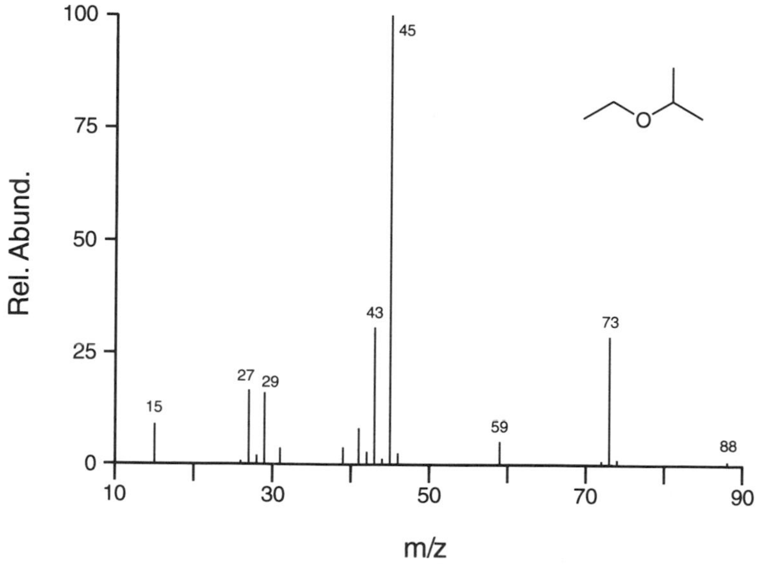

**Figure 5.21.** Mass spectrum of ethyl isopropyl ether. Ions resulting from rearrangement of initially formed α-cleavage ions are more intense in the spectra of aliphatic alcohols and ethers.

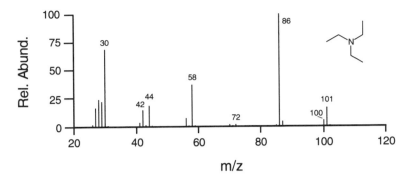

**Figure 5.22.** Mass spectrum of triethylamine. Secondary losses from the initial α-cleavage ions at *m/z* 100 and 86 account for the ions at *m/z* 72, 58, and 44. The ion at *m/z* 30 results from an additional secondary loss from *m/z* 58.

The rearrangement in Eq. (5.16) accounts for the small ion at *m/z* 59, which is otherwise inexplicable since cleavage of the carbon–heteroatom bond is not observed in these compounds.

Some tertiary amines have two alkyl groups that can undergo this elimination. For instance, the spectrum of triethylamine (Figure 5.22) shows the expected loss of a methyl group to give the base peak at *m/z* 86. However, this ion contains two alkyl groups meeting the criterion for rearrangement. Elimination of ethylene from one of these groups leads to the fairly intense ion at *m/z* 58, which in turn loses ethylene from the final ethyl group to produce the ion at *m/z* 30:

$$\text{(5.19)}$$

| m/z 101 | m/z 86 | m/z 58 | m/z 30 |

## 5.6. ELIMINATION IN SOME AROMATIC ESTERS AND AMIDES

In compounds containing aliphatic oxygen or nitrogen, α-cleavage may account for much, if not nearly all, of the important initial fragmentation in the molecule. This is true even for compounds in which the heteroatom is located near an aromatic ring. For structures in which the heteroatom is attached directly to the ring, however, an alternate mode of fragmentation can occur—olefin elimination via a four-membered cyclic intermediate.

The spectrum of acetylacetaminophen, the *O*-acetylated derivative of analgesics such as Tylenol, illustrates this well (Figure 5.23). Although the ion at *m/z* 43 arises from α-cleavage at either carbonyl group, there are no α-cleavage ions corresponding to the loss of methyl [(M − 15)⁺; *m/z* 178] or acetyl [(M − 43)⁺; *m/z* 150]. Instead the only two high-mass ions other than the molecular ion occur at *m/z* 151 [(M

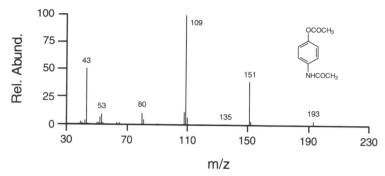

**Figure 5.23.** Mass spectrum of acetylacetaminophen. Acylated derivatives of phenols and aromatic amines lose ketene rather than undergo α-cleavage at the carbonyl group.

$- 42)^+]$ and $m/z$ 109 $[(M - 42–42)^+]$, produced by the sequential cyclic losses of ketene ($CH_2=C=O$):

$$(5.20)$$

m/z 193          m/z 151                     m/z 109

Once again, these fragment ions have stabilities similar to those of their parent ions and form by four-center rearrangement of hydrogen and elimination of a small molecule containing a newly formed π-bond. In this case the product ions are both "molecular ions" of aromatic compounds that have substantial stability in their own right. This type of fragmentation also explains the behavior of phenylalkylethers such as ethoxybenzene, which loses ethylene ($m/z$ 94; 100%) to the near exclusion of all other fragmentation:

$$- CH_2CH_2 \qquad (5.21)$$

m/z 122                          m/z 94

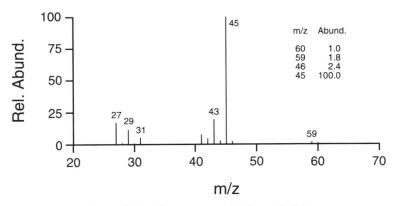

**Figure 5.24.** Mass spectrum for Example 5.1.

## 5.7. EXAMPLES AND PROBLEMS

***Examples 5.1 and 5.2.*** Identify the compounds whose spectra are shown in Figures 5.24 and 5.25.

*Answer 5.1.* The ion at *m/z* 59 cannot be the molecular ion and also lose 14 daltons to produce *m/z* 45; the tiny ion at *m/z* 60 seems a more likely candidate. The isotope cluster for the base peak at *m/z* 45 shows that it contains two carbons but indicates no sign of the presence of silicon or sulfur. As such, it is indicative of α-cleavage from an aliphatic ether or alcohol (Table 4.2). Only three structures are possible: *n*-propanol, isopropanol, and methylethylether. The latter two compounds are consistent with the spectrum (*n*-propanol should produce a base peak of *m/z* 31 by α-cleavage). Although it is tempting to speculate on which of the two remaining compounds produces this spectrum, only comparison with known standards will settle the issue unambiguously. In actuality, the spectra of these two compounds are very similar, although methylethylether shows a larger molecular ion, an enhanced $(M - 1)^+$ ion

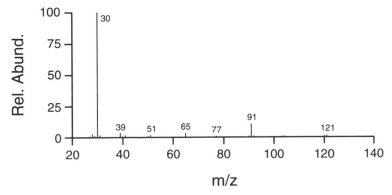

**Figure 5.25.** Mass spectrum for Example 5.2.

from the five hydrogens on the α carbons, and a more intense *m/z* 31 ion from secondary rearrangement of one of the *m/z* 59 ions:

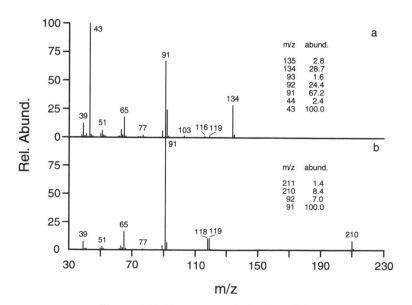

$$\underset{m/z\ 45}{\text{⊕O}}\xleftarrow[a]{-CH_3^{\bullet}}\underset{m/z\ 60}{\overset{b}{H}\overset{a}{O}}\xrightarrow[b]{-H^{\bullet}}\underset{m/z\ 59}{H_2C=\overset{⊕}{O}}\xrightarrow{-CH_2CH_2}\underset{m/z\ 31}{H_2C=\overset{⊕}{O}-H}$$ (5.22)

Isopropanol, on the other hand, has a more intense *m/z* 43 ion from the isopropyl group itself (isopropanol; $CH_3CHOHCH_3$).

*Answer 5.2.* The base peak at *m/z* 30 and the apparently odd molecular weight strongly suggest an aliphatic amine. This is accompanied by a weak, but significant, benzyl low mass ion series (*m/z* 91, 65, and 39; they are about the only other ions in the spectrum). The combination of benzyl and $CH_2=NH_2\oplus$ (*m/z* 30) gives β-phenethylamine (ϕ-$CH_2CH_2NH_2$), which has a molecular weight of 121, consistent with the molecular ion. You should convince yourself that no other isomeric structure can produce the intense *m/z* 30 ion by α-cleavage (β-phenethylamine):

$$\underset{m/z\ 30}{CH_2=\overset{⊕}{N}H_2}\xleftarrow[a]{-\phi CH_2^{\bullet}}\underset{m/z\ 121}{\overset{a}{\underset{b}{\bigcirc}}\overset{+}{N}H_2}\xrightarrow[b]{-CH_2=\overset{\bullet}{N}H_2}\underset{m/z\ 91}{\overset{⊕}{\bigcirc}}$$ (5.23)

**Figure 5.26.** Mass spectra for Problem 5.5.

***Problem 5.5.*** The compounds whose spectra are shown in Figure 5.26 were recovered from a clandestine drug lab in which methamphetamine was being manufactured. Use this information, as well as your mass spectral interpretation skills, to identify these compounds.

***Problem 5.6.*** Figure 5.27 contains the spectra of the seven $C_4H_{10}O$ isomers. Draw the structures for each of these compounds and predict important modes of fragmentation for each, then match the structures with the spectra.

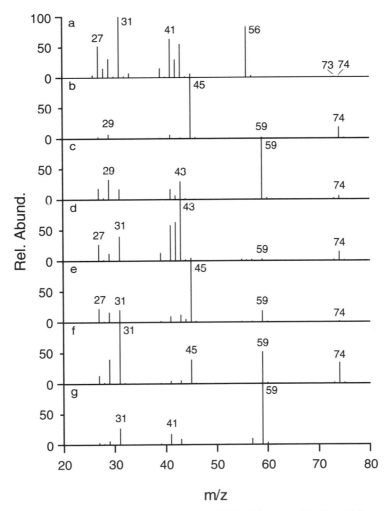

**Figure 5.27.** Mass spectra of the seven $C_4H_{10}O$ isomers (Problem 5.6).

**Problem 5.7.** Below are the structures of 10 compounds that might be encountered during the analysis of illicit drug samples. Identify in each structure the primary sites for α-cleavage. Then account for formation of the base peak, or other peaks as indicated, in each spectrum. Ions marked with asterisks are by far the most intense in the spectrum.

a. Benzocaine (*m/z* 120):

b. 1,3-diphenyl-2-aminopropane (by product of methamphetamine synthesis) (*m/z* 134*, 91):

c. Mepivacaine (*m/z* 98):

d. Anileridine (*m/z* 246*):

e. Doxepin (*m/z* 58*):

f. Flurazepam (*m/z* 86*):

g. *N,N*-diethylamphetamine (*m/z* 100, 91, 72):

h. Methadone (*m/z* 72*):

i. Cocaine (*m/z* 272, 105):

j. α-methylfentanyl ("China White;" *m/z* 259):

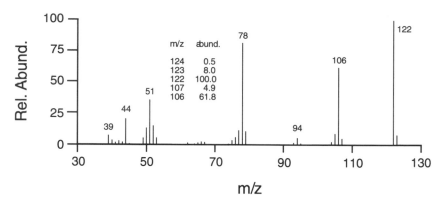

**Figure 5.28.** Mass spectrum for Problem 5.8.

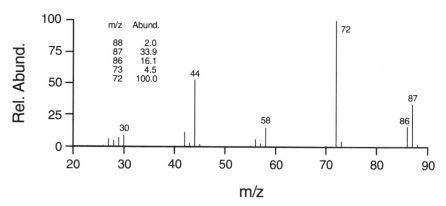

**Figure 5.29.** Mass spectrum for Problem 5.9.

*Problems 5.8. and 5.9.* Identify the compounds for which spectra are shown in Figures 5.28 and 5.29.

## REFERENCES

*Forensic and Analytical Chemistry of Clandestine Phenethylamines,* CND Analytical, Inc., Auburn, AL, 1994, pp. 31–50.

R.J. Renton, J.S. Cowie, and M.C.H. Oon, *Forensic Sci. Int.,* **60,** 189–202 (1993).

# CHAPTER 6

# IMPORTANT MASS SPECTRAL REARRANGEMENTS

## 6.1. INTRODUCTION

Fragmentations involving rearrangement of the atoms and electron density of parent ions were introduced in Chapters 4 and 5. In each of these fragmentations the rearrangements were driven by elimination of a small neutral molecule and formation of an ion that had stability similar to that of the starting ion.

In this chapter, we will examine four specific types of mass spectral rearrangements—the γ-hydrogen rearrangement, cyclohexanone-type rearrangement, retro Diels–Alder fragmentation, and double-hydrogen rearrangement. These fragmentations provide specific structural information about parts of the molecule under investigation simply because they do not occur unless the participating atoms are positioned so that efficient rearrangement of electron density can take place.

Many of the rearrangements discussed previously involved four-center intermediates or transition states, although the loss of water in aliphatic alcohols began with the six-center transfer of hydrogen. The size of the "rings" formed as intermediates for these rearrangements is not accidental. We might expect rearrangements involving rings of three or four atoms to proceed with facility simply because the atoms are in close proximity to one another. Rings involving six atoms are also easy to form for entropic and structural reasons and are preferred intermediates for many rearrangements in organic molecules. Many of the rearrangements encountered in mass spectrometry proceed via three-, four-, or six-center intermediates; those involving five- or seven-center intermediates occur less frequently. Three of the rearrangements discussed in this chapter involve six-center intermediates.

## 6.2. GAMMA-HYDROGEN REARRANGEMENT

### 6.2.1. McLafferty-Type Rearrangement

Figure 6.1 summarizes a rearrangement involving the migration of a hydrogen atom attached to a carbon four atoms removed from the initial ionization site and the subsequent rearrangement of electron density leading to the loss of an olefin—a fragmentation known as the $\gamma$-hydrogen rearrangement. Within this general framework several combinations are possible. When the X group in Figure 6.1 is oxygen (i.e., for carbonyl compounds), this fragmentation is called the "McLafferty-type" rearrangement, named after Fred McLafferty of Cornell University, a pioneer in mass spectral interpretation.

The individual steps in the mechanism shown in Figure 6.1 are all analogous to processes discussed previously. Initial ionization occurs at the heteroatom, and hydrogen transfer occurs via a six-center intermediate. Redistribution of electron density and ejection of an olefin leads to formation of an ion whose stability is similar to that of the starting ion, and the fragmentation is driven by the energy gained in forming a new $\pi$-bond in the olefinic product. Since this mechanism involves several steps, it is often abbreviated in the shorthand form shown in Eq. (6.1) using full arrows.

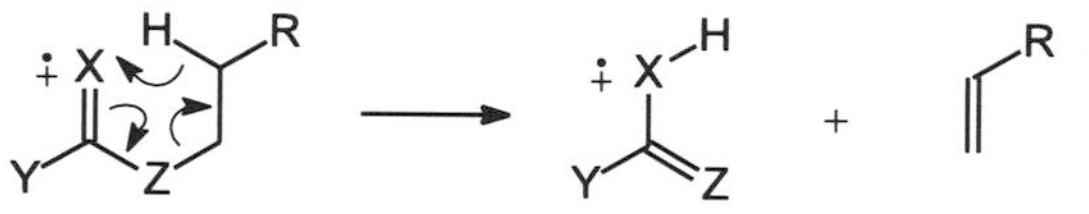

$$(6.1)$$

It must be remembered, however, that this shorthand notation oversimplifies what occurs during this rearrangement and may gloss over the impact of structural differences that determine the relative propensities of different compounds to undergo it.

The position of the hydrogen in this rearrangement is critical; the reaction does not proceed unless a hydrogen is available on the $\gamma$-carbon. Consider the two spectra in

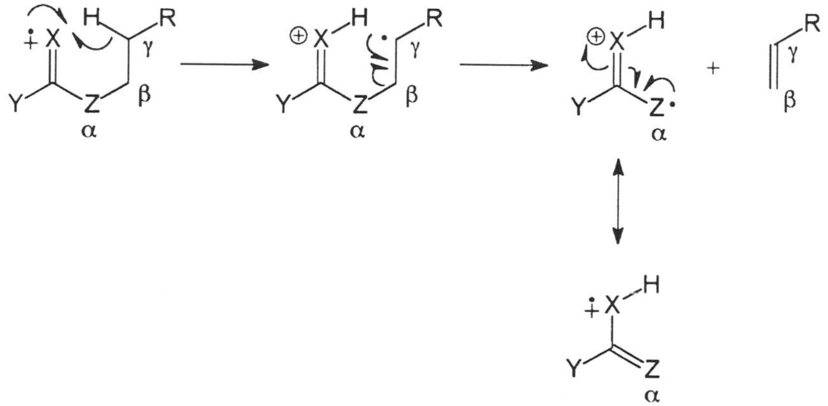

**Figure 6.1.** Generalized representation of the $\gamma$-hydrogen rearrangement, where X = O, $CR_2$, Y = R, OR, $NR_2$, and Z = $CR_2$, O.

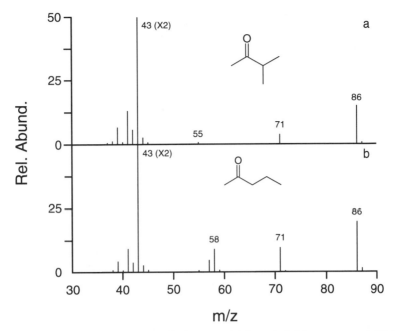

**Figure 6.2.** Mass spectra of two aliphatic ketones: (*a*) 3-methyl-2-butanone and (*b*) 2-pentanone. The ion at *m/z* 58 in the spectrum of 2-pentanone is the result of γ-hydrogen rearrangement.

Figure 6.2. 3-Methyl-2-butanone (Figure 6.2*a*) undergoes α-cleavage as expected to give fragment ions at *m/z* 71 and 43 (cleavage on either side of the carbonyl group), with the ion at *m/z* 43 being more abundant because of the difference in stability between isopropyl and methyl radical. The ions at *m/z* 41 and 39 probably come from fragmentation of the isopropyl ion to give the allyl ($CH_2=CH-CH_2\oplus$) and cyclopropenium ($C_3H_3\oplus$) ions, respectively. These are the only fragment ions of any consequence in this spectrum.

The spectrum of 2-pentanone (Figure 6.2*b*), on the other hand, shows, in addition to the ions from α-cleavage, a prominent ion at *m/z* 58 that does not occur in the spectrum of 3-methyl-2-butanone. The mass of this ion stands out because the masses of all of the other important fragment ions in both spectra occur at odd mass. Since neither of these compounds contains nitrogen, the odd-nitrogen rule predicts that any even-mass ion in this spectrum will also be odd-electron. This means that *m/z* 58 is a likely candidate for formation by a rearrangement process.

The structure of this ion and the mechanism for its formation cannot be determined from this information alone. The loss from the molecular ion is 28 daltons, most likely either carbon monoxide or ethylene; both are possible in this molecule. High-resolution mass spectrometry tells us that the loss is ethylene, not carbon monoxide, but does not define what parts of the molecule are lost in forming the rearranged ion.

Instead, a method of labeling atoms is needed to do this. Deuterium, an isotope of hydrogen having an extra neutron, has an atomic mass of 2. Replacement of some of

**Table 6.1. Following the McLafferty Rearrangement in 2-Pentanone**

| Compound | M$^{+\cdot}$ | PrCO$^+$ | CH$_3$CO$^+$ | Rearrangement Ion |
|---|---|---|---|---|
| CH$_3$COCH$_2$CH$_2$CH$_3$ | 86 | 71 | 43 | 58 |
| CD$_3$COCH$_2$CH$_2$CH$_3$ | 89 | 71 | 46 | 61 ($\therefore$ all D retained) |
| CH$_3$COCD$_2$CH$_2$CH$_3$ | 88 | 73 | 43 | 60 ($\therefore$ all D retained) |
| CH$_3$COCH$_2$CD$_2$CH$_3$ | 88 | 73 | 43 | 58 ($\therefore$ all D lost) |
| CH$_3$COCH$_2$CH$_2$CD$_3$ | 89 | 74 | 43 | 59 ($\therefore$ 2 D lost, 1 D retained) |

the hydrogen atoms in a molecule by deuterium increases the molecular weight by 1 dalton for each hydrogen replaced—a mass shift readily apparent in mass spectrometry. If one or more hydrogens at a specific location in a molecule are replaced with deuterium by reaction with a reagent such as deuterium oxide (D$_2$O; "heavy water"), that site can be followed through various fragmentations by studying how ions in the spectrum containing that fragment shift in mass. In this case, four deuterated derivatives of 2-pentanone were synthesized unambiguously. The syntheses of these compounds were planned carefully since exchange of hydrogen for deuterium in 2-pentanone itself by reaction with most deuterated reagents would lead to an inseparable mixture of compounds. In one derivative all of the hydrogens on carbon 1 were replaced with deuterium, in another all of the hydrogens on carbon 3, and so forth. Important peaks in the mass spectra of each of these compounds, and of the parent 2-pentanone, are listed in Table 6.1.

In the spectrum of the first derivative, 1-$d_3$-2-pentanone, the molecular ion, instead of being found at m/z 86 for $^{12}$C$_5$H$_{10}$$^{16}$O, is found at m/z 89 for $^{12}$C$_5$D$_3$H$_7$$^{16}$O. The ion at m/z 71 in the 2-pentanone spectrum, due to loss of the methyl attached to the carbonyl group, also occurs at m/z 71 in this spectrum, since carbon 1 and all of its hydrogens (deuteriums) are lost in this fragmentation. However, the acetyl ion that occurs at m/z 43 in the spectrum of 2-pentanone moves to m/z 46 in the 1-$d_3$-2-pentanone spectrum since all of the deuteriums remain with the positive charge in this fragmentation. Finally, the rearrangement ion at m/z 58 moves to m/z 61 in the 1-$d_3$-2-pentanone spectrum, indicating that, whatever the mechanism, carbon 1 and its attached hydrogens are still present in the resulting ion (barring, of course, the possibility that carbon 1 is lost, but *all* of its hydrogens are rearranged to some other position; other experiments would be necessary to show that this is not the case).

In the spectrum of 3-$d_2$-2-pentanone the molecular ion is found at m/z 88 with the addition of two deuteriums, the ion arising from α-cleavage loss of methyl moves to m/z 73 (since the deuteriums now remain with the positive charge), and the acetyl ion remains at m/z 43. The ion that occurs at m/z 58 in the spectrum of parent compound moves to m/z 60 in this spectrum, indicating that carbon 3 and both of its deuteriums are retained in the fragmentation.

The spectrum of 4-$d_2$-2-pentanone is similar, except for the rearrangement ion. In this case it remains at m/z 58, consistent with the *loss* of carbon 4 and its attached hydrogens. This accounts for half of the CH$_2$=CH$_2$ lost in this reaction.

Finally, the rearrangement ion moves to m/z 59 in the spectrum of 5-$d_3$-2-pentanone, a change of only 1 dalton from the parent compound. This means that *one* of the three deuterium atoms located on carbon 5 is *retained* in the fragment ion. Since the 58 daltons in the fragment ion have already been accounted for as $C_3H_6O$, the remaining two deuteriums *and* carbon 5 must be lost as the other half of the ethylene molecule.

These findings are completely compatible with the mechanism in Figure 6.1: one hydrogen from the γ-carbon (carbon 5, in this case) migrates to that part of the molecule containing carbon atoms 1, 2, and 3, while the ß- and γ-carbons (carbons 4 and 5) and four of the five groups attached to them are lost. As long as the γ-carbon has at least one rearrangeable hydrogen, the ß- and γ-carbons can have a variety of substituents. Generalizing beyond alkyl groups is dangerous, however, since the electronic structure of some molecules may permit other fragmentation processes to compete favorably. 3-Methyl-2-butanone (Figure 6.2a) does not have a γ-carbon and does not undergo the rearrangement.

The McLafferty rearrangement is not limited to aliphatic ketones. Aliphatic carboxylic acids and their derivatives also undergo this fragmentation, which sometimes produces the most abundant ions in the spectrum [Figure 6.3 and Eq. (6.2)]:

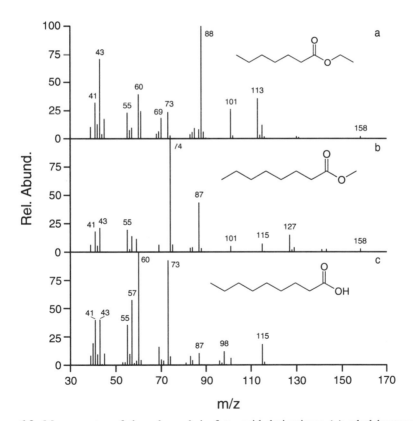

**Figure 6.3.** Mass spectra of three long-chain fatty acid derivatives: (*a*) ethyl heptanoate, (*b*) methyl octanoate, and (*c*) nonanoic acid. In each case the McLafferty rearrangement ion is the base peak.

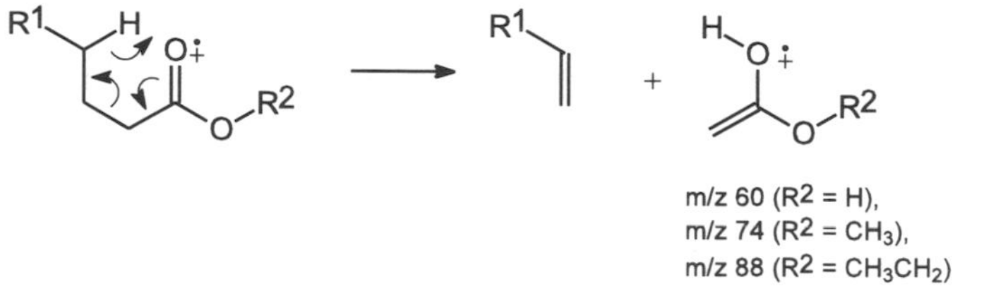

$$ \text{(6.2)} $$

m/z 60 (R2 = H),
m/z 74 (R2 = CH₃),
m/z 88 (R2 = CH₃CH₂)

Nor is the rearrangement limited to the aliphatic portion of the molecule (Z=CH$_2$ in Figure 6.1). Butyl palmitate, for example, exhibits two peaks of nearly equal intensity due to McLafferty rearrangements involving the ester oxygen [Z=O in Figure 6.1; Eq. (6.3)] and the aliphatic chain, respectively:

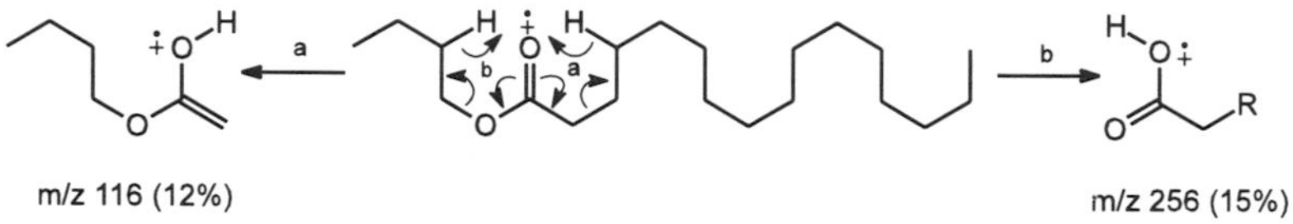

$$ \text{(6.3)} $$

m/z 116 (12%)                                      m/z 256 (15%)

For compounds in which the McLafferty rearrangement ion gains aromatic resonance energy unavailable to the parent ion, the rearrangement ions may be strikingly intense, as in the spectra of the barbiturates [Figure 6.4 and Eq. (6.4)]:

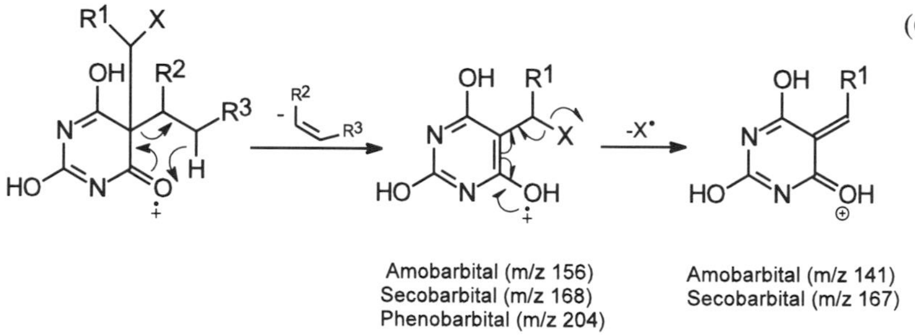

$$ \text{(6.4)} $$

Amobarbital (m/z 156)            Amobarbital (m/z 141)
Secobarbital (m/z 168)           Secobarbital (m/z 167)
Phenobarbital (m/z 204)

### 6.2.2. Gamma-Hydrogen Rearrangement in Alkylbenzenes

As expected, the alkylbenzenes whose spectra are shown in Figure 6.5 give aromatic and benzylic low mass ion series and undergo benzylic cleavage to give a base peak at m/z 91. Closer scrutiny, however, reveals that the m/z 92 ion in the spectrum of n-butylbenzene (Figure 6.5c) is much too large to be due to the natural isotopic abundance of $^{13}$C. In fact, the m/z 92 ion even in the spectrum of n-propylbenzene has an intensity of about 10%, slightly larger than the calculated 7.7%. Although this difference might easily be overlooked in the case of n-propylbenzene, intensity measurement errors cannot possibly account for the size of this ion in the butylbenzene spectrum.

Since m/z 92 is an odd-electron ion (even molecular weight and no nitrogens), its formation by some sort of rearrangement process seems likely. If one of the "double

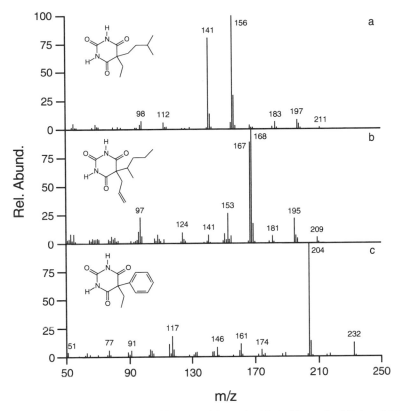

**Figure 6.4.** Mass spectra of three barbiturates: (*a*) amobarbital, (*b*) secobarbital, and (*c*) phenobarbital. The McLafferty rearrangement ions are the base peaks in these spectra.

bonds" in the aromatic ring undergoes the initial ionization (there are no heteroatoms in these molecules), this nucleophilic site could attract an appropriately situated hydrogen to migrate to one of the ring positions. Given the fact that ethylbenzene (Figure 6.5*a*) does not undergo this fragmentation, the hydrogen attached to the carbon γ to the ring provides a convenient source:

(6.5)

Table 6.2 lists the relative sizes of the *m/z* 92 and 91 ions in the spectra of a variety of alkylbenzenes. (For those compounds having an additional methyl group in the benzylic position, the benzylic cleavage and γ-hydrogen rearrangement ions occur at *m/z* 105 and 106, respectively). The data in this table support our hypothesis about

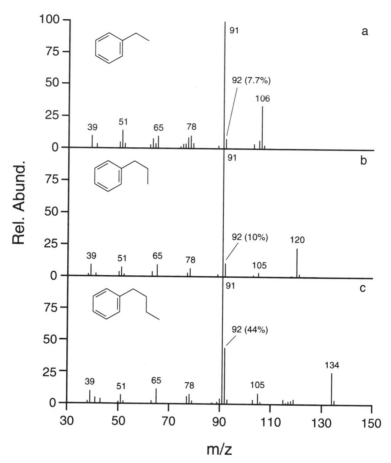

**Figure 6.5.** Mass spectra of three alkylbenzenes: (*a*) ethylbenzene, (*b*) *n*-propylbenzene, and (*c*) *n*-butylbenzene. The intensity of *m/z* 92 increases with the facility of γ-hydrogen rearrangement.

the role of the γ-hydrogen in this fragmentation. First, compounds with no γ-hydrogens do not undergo this rearrangement. For example, 1-phenyl-3,3-dimethylbutane (Table 6.2, bottom left) produces an *m/z* 92 ion comparable to that expected from isotopic abundances alone. On the other hand, in the spectrum of 1-phenyl-2,2-dimethylpropane (neopentylbenzene; Table 6.2, bottom right) *m/z* 92 is over one and a half times the size of the *m/z* 91 ion, reflecting the fact that this compound has nine γ-hydrogens that can rearrange.

One variable affecting the activation energy necessary for this rearrangement appears to be the strength of the incipient double bond in the resulting olefin—compounds producing substituted olefins undergo the rearrangement more readily than those that do not. For example, propylbenzene, which generates only ethylene, shows minimal evidence for fragmentation via γ-hydrogen rearrangement, but 1-phenyl-2-

**Table 6.2.  γ-Hydrogen Rearrangements in Various Alkylbenzenes**

| m/z 92/91 | m/z 92/91 | m/z 92/91 |
|---|---|---|
| 0.08 | 0.6 | 1.2 |
| 0.10 | 0.7 | 1.3 |
| 0.1[a] | 0.8 | 1.6 |
| 0.1[a] | 0.6 | |
| 0.1 | 1.0[a] | |

[a] m/z 106/105.

methylbutane (Table 6.2, second from top right) undergoes this fragmentation more easily than does the isomeric n-pentylbenzene (Table 6.2, second from top middle), apparently because the olefin formed has two attached alkyl groups rather than one.

The presence of additional alkyl groups on the benzylic carbon, however, increases the relative stability of the benzylic ion. Even in cases where several hydrogens can rearrange (1-phenyl-1,2-dimethylpropane; Table 6.2, third from top left), benzylic cleavage is the preferred fragmentation. Only in the case of 1-phenyl-1,2,2-trimethyl-propane (Table 6.2, bottom middle) does the statistical probability of having many migratable hydrogens and the stability of the resulting olefin allow γ-hydrogen re-arrangement to compete favorably with benzylic cleavage.

### 6.2.3. Gamma-Hydrogen Rearrangement in Aromatic Compounds with Heteroatoms

An alkylbenzene-type γ-hydrogen rearrangement also may occur in aromatic com-pounds containing heteroatoms. For example, the m/z 258 ion in the spectrum of $\Delta^9$-tetrahydrocannabinol (THC, the biologically active component of marijuana) arises from just this type of fragmentation [Figure 6.6a and Eq. (6.6)]:

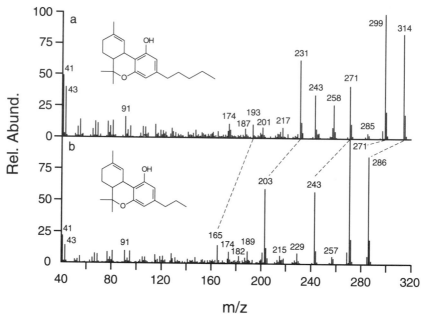

**Figure 6.6.** Mass spectra of (*a*) Δ⁹-tetrahydrocannabinol and (*b*) tetrahydrocannabivarin. Note the correspondence between peaks in (*a*) that occur at masses 28 daltons lower in (*b*). All of these ions should contain the entire alkyl side chain. The ion at *m/z* 258 in (*a*), due to γ-hydrogen rearrangement, is not reproduced in (*b*).

In this case initial ionization occurs at one of the heteroatoms, and the "γ-hydrogen rearrangement" involves a more substantial redistribution of the π-electron density— "γ-hydrogen rearrangement by long distance," if you will.

Support for the mechanism shown in Eq. (6.6) comes from the spectrum of $\Delta^9$-tetrahydrocannabivarin, a homolog of THC having a propyl side chain (Figure 6.6b). In this case the ion at $m/z$ 258 is much less intense than in the THC spectrum, consistent with the decreased propensity for propylbenzenes to undergo this rearrangement (Table 6.2). The intense ion at $m/z$ 243 in this spectrum corresponds to the $m/z$ 271 ion in the spectrum of THC and appears to be produced by loss of the entire *gem*-dimethyl group and its attached carbon (see Problem 7.3). A summary of some of the major fragmentations of THC is given in Figure 6.7.

Because of its inherent basicity and electronegativity, nitrogen in the aromatic ring enhances the preference for $\gamma$-hydrogen rearrangement. Consider the spectra of 4- and 2-propylpyridine (Figure 6.8). Both compounds readily undergo $\gamma$-hydrogen rearrangement to produce ions at $m/z$ 93. 2-Propylpyridine does this more easily, espe-

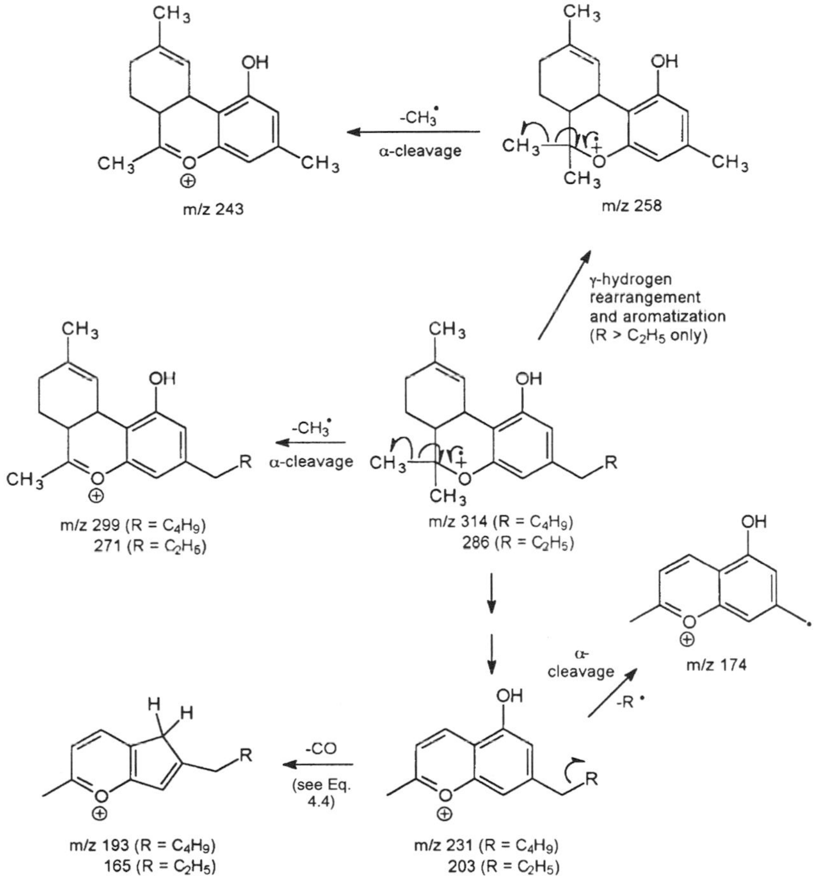

**Figure 6.7.** Some fragmentations of $\Delta^9$-tetrahydrocannabinol. $\alpha$-Cleavage from the molecular ion and $m/z$ 258 lead to the ions at $m/z$ 299 and 243, respectively, while the benzylic cleavage from $m/z$ 231 to produce $m/z$ 174 is analogous to that shown in Eq. (5.11) and loss of CO from $m/z$ 231 mirrors that seen in Eq. (4.4). Formation of $m/z$ 231 is not straightforward. (Adapted with permission from Smith, 1997; copyright ASTM).

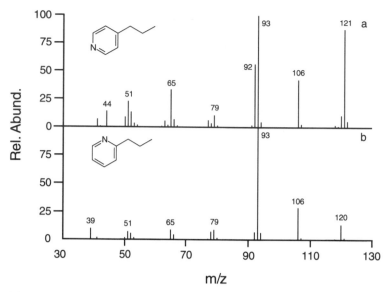

**Figure 6.8.** Mass spectra of (*a*) 4-propylpyridine and (*b*) 2-propylpyridine. These spectra differ substantially because of interactions that are possible between the propyl group and the adjacent nitrogen in the 2-propyl isomer ("ortho effect").

cially compared to fragmentation by benzylic cleavage (*m/z* 92), because the migrating hydrogen is attracted directly to the initial ionization site [Eqs. (6.7) and (6.8)]:

Indeed, this pathway requires so little energy in 2-propylpyridine that the molecular ion is extremely weak in intensity.

The difference in fragmentation patterns observed for these two compounds is an example of the "*ortho* effect," a term denoting the tendency for aromatic compounds having substituents located *ortho* to one another to undergo reactions that are different from those where the same substituents are *meta* or *para* to one another. *Ortho* substituents can interact with one another via cyclic intermediates in ways that *meta* and *para* substituents cannot. For example, the facile loss of hydrogen radical by 2-

propylpyridine can be rationalized by a mechanism in which a ring is formed, the positive charge is stabilized on nitrogen, and the two radical sites are neutralized:

(6.9)

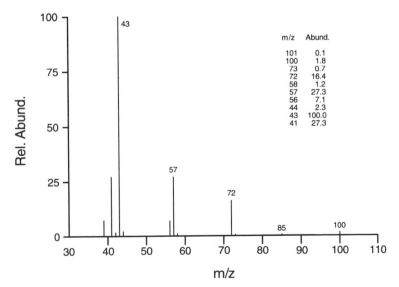

The 4-propyl isomer, on the other hand, cannot easily shift electron density to neutralize the radical site in this fragmentation, so that other fragmentations prevail:

(6.10)

**Problem 6.1.** Identify the compound that produced the spectrum shown in Figure 6.9.

## 6.3. CYCLOHEXANONE-TYPE REARRANGEMENT

The spectrum of cyclohexanone (Figure 6.10) has some surprising features. First is the loss of a methyl group to give the ion at $m/z$ 83, although this loss seems some-

| m/z | Abund. |
|-----|--------|
| 101 | 0.1 |
| 100 | 1.8 |
| 73 | 0.7 |
| 72 | 16.4 |
| 58 | 1.2 |
| 57 | 27.3 |
| 56 | 7.1 |
| 44 | 2.3 |
| 43 | 100.0 |
| 41 | 27.3 |

**Figure 6.9.** Mass spectrum for Problem 6.1.

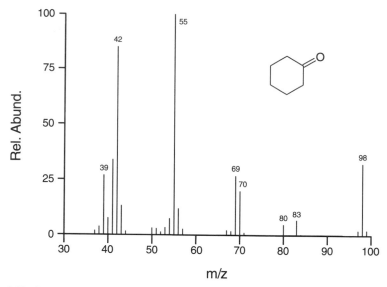

**Figure 6.10.** Mass spectrum of cyclohexanone. In order for this molecule to fragment, the ring must first open.

what more reasonable here than it does in the case of benzene. In addition, the losses of water (*m/z* 80), ethylene and carbon monoxide (*m/z* 70; as it turns out, *both* are lost), ethyl radical (*m/z* 69), and either propyl or acetyl radical (*m/z* 55) hardly seem intuitive.

Insight into the fragmentations of cyclohexanone can be gained by considering first what we expect this molecule to do. Cyclohexanone is an unsubstituted carbonyl compound in which the γ-hydrogens are sterically restricted by the ring structure and therefore cannot undergo the McLafferty rearrangement. Hence, the only pathway left after initial ionization at the carbonyl oxygen is α-cleavage on either side of the carbonyl group. However, because the carbonyl carbon is part of the ring, this cleavage involves no loss of mass. Other fragmentations must occur before ions are formed that are detectable by mass spectrometry.

As with the McLafferty rearrangement, determining a mechanism for formation of the base peak at *m/z* 55 involved studying the mass spectra of labeled derivatives— in this case, the three deuterated derivatives shown in Table 6.3. Deuterium was used in these studies, rather than alkyl derivatives, for example, because the ions formed from deuterated compounds would have essentially the same stability as those of the parent compound. In contrast, competing fragmentations in alkylated derivatives could obscure the effects being studied. As before (Section 6.2.1), we assume that, if all of the deuteriums are lost from a carbon atom during fragmentation, that carbon is lost as well.

In the spectrum of 2,2,6,6-tetradeuteriocyclohexanone the base peak is found at *m/z* 56—a shift of 1 dalton, indicating that three of the four deuteriums are lost in the fragmentation leading to its formation. At this point it cannot be determined if one or both of the carbons containing these deuteriums is lost as well. The spectrum of 3,3,5,5-tetradeuteriocyclohexanone has a base peak at *m/z* 57, consistent with the loss

**Table 6.3. Cyclohexanone Rearrangement—Deuterium Labeling Studies**

| Base Peak | | Base Peak |
|---|---|---|
| | 55 | |
| | | 57 (2 of 4 D's retained) |
| | 56 (only 1 of 4 D's retained) | |
| | | 55 (both D's lost) |

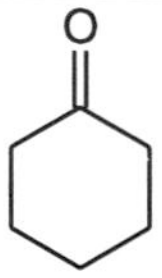

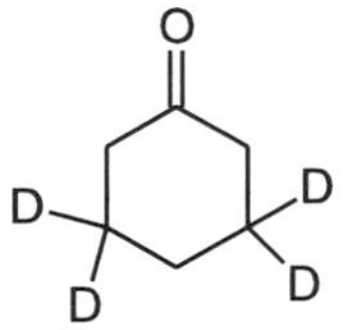

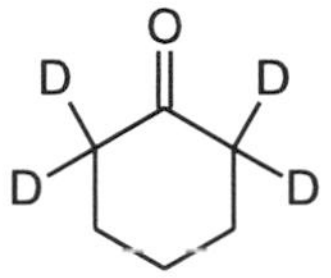

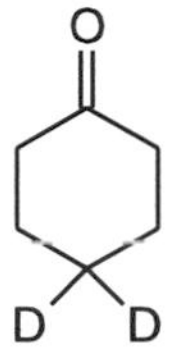

of two of the four deuteriums. This indicates that either carbon 3 or 5 (it does not matter which one—they are equivalent by symmetry) and its attached deuteriums are lost from this compound. Finally, the base peak in the spectrum of 4,4-dideuteriocyclohexanone occurs at $m/z$ 55, signaling the loss of carbon 4 and its attached deuteriums.

The data in Table 6.3 thus describe the following scenario: In the formation of the base peak, carbon 6 (or carbon 2; they are equivalent) and its hydrogens, carbon 5 and its hydrogens, carbon 4 and its hydrogens, and one of the hydrogens from carbon 2 are lost—a total of three carbons and seven hydrogens, that is, a propyl group. Carbons 4, 5, and 6 and their hydrogens are all contiguous, but loss of the hydrogen from carbon 2 must still be explained. By analogy with other fragmentations, migration of a hydrogen radical from carbon 2 to the primary radical site on carbon 6 in the initially formed α-cleavage ion can proceed via a six-membered cyclic intermediate:

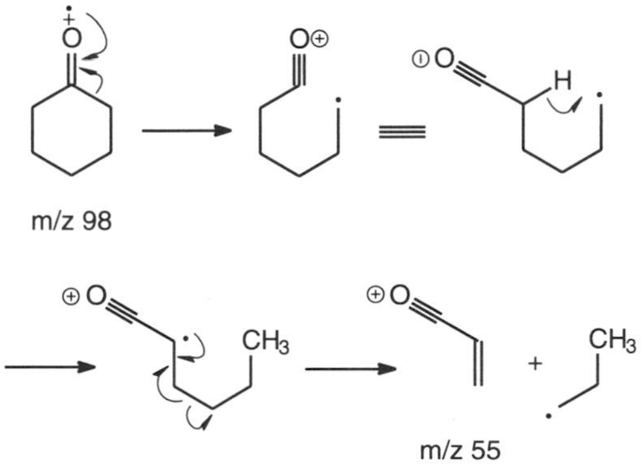

(6.11)

m/z 98

m/z 55

Since the incipient radical site on carbon 2 is not well tolerated due to the neighboring positively charged carbonyl group, the bond between carbons 3 and 4 breaks easily to form a propyl radical and an α-cleavage-type ion that is stabilized by conjugation with the newly formed double bond. All of the steps in this fragmentation are similar to reactions discussed previously in other contexts.

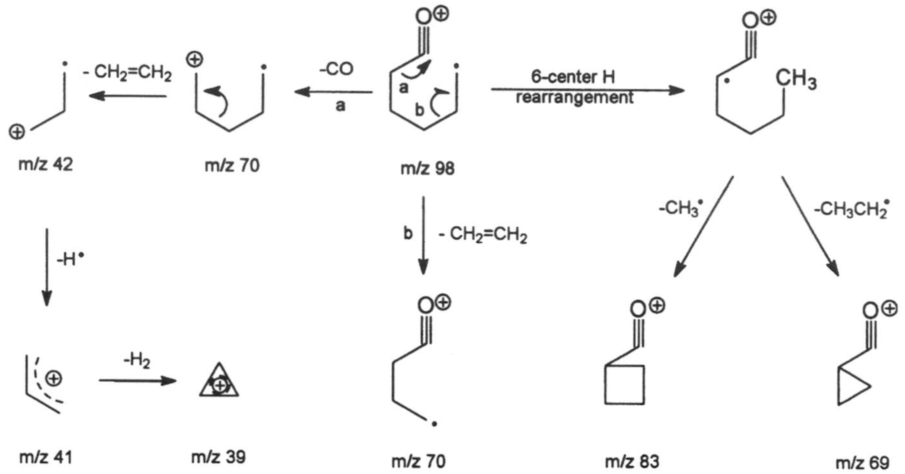

**Figure 6.11.** Other fragmentations of cyclohexanone.

The remaining major fragmentations of cyclohexanone are consistent with this mechanism (Figure 6.11). At the right of this figure, the radical formed after hydrogen rearrangement also can lose either methyl or ethyl radical to produce the ions at $m/z$ 83 and 69, respectively. Either the substantial energy needed to form the small rings shown here or the electronic factors associated with formation of the corresponding diradicals are significant enough that these ions are less intense than that resulting from the cyclohexanone rearrangement.

The radical ion formed from $\alpha$-cleavage loses carbon monoxide to give one of the two ions at $m/z$ 70 (both ions are seen at high resolution), which in turn loses ethylene like other primary aliphatic ions to produce the ion at $m/z$ 42 [Eq. (4.2)]. Hydrogen radical loss from $m/z$ 42 leads to the allyl ion ($m/z$ 41), and further loss of a hydrogen molecule forms the aromatic cyclopropenium ion ($m/z$ 39). An alternate fragmentation of the initially formed radical ion involves direct loss of ethylene to form a second ion at $m/z$ 70, which has a stability similar to that of the initial intermediate. Once again, formation of a small unsaturated molecule drives this fragmentation.

The cyclohexanone-type rearrangement is applicable to other cyclohexane derivatives. The initial step in this fragmentation is $\alpha$-cleavage and, since the remaining steps do not involve direct involvement of the carbonyl oxygen per se, any functional group capable of causing $\alpha$-cleavage should also initiate the entire rearrangement (Figure 6.12). The spectra of two other cyclohexane derivatives, methoxycyclohexane and dimethylaminocyclohexane (Figure 6.13), show base peaks resulting from cyclohexanone-type rearrangements:

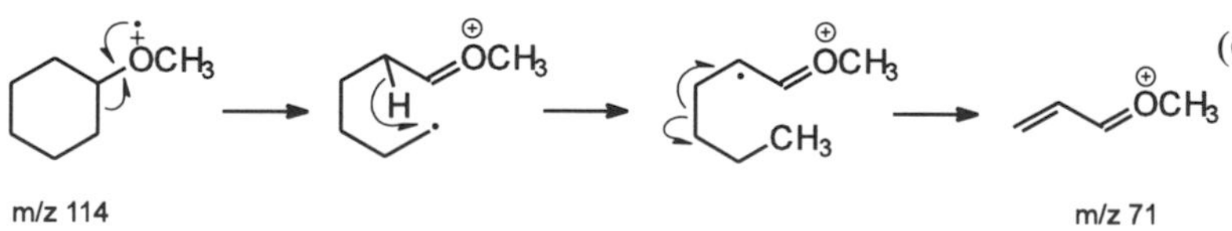

(6.12)

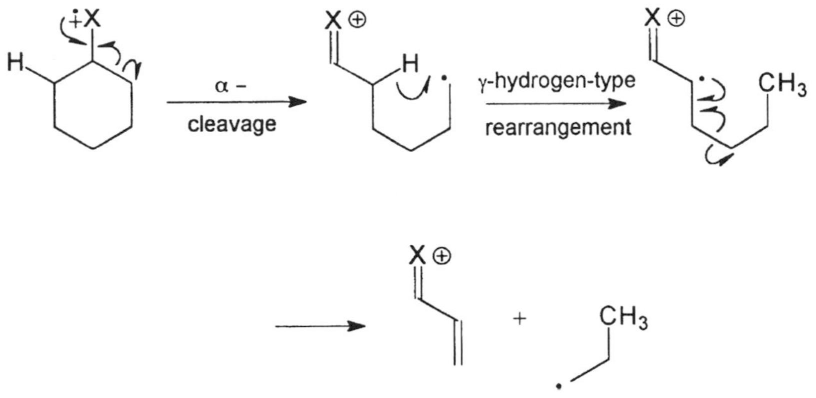

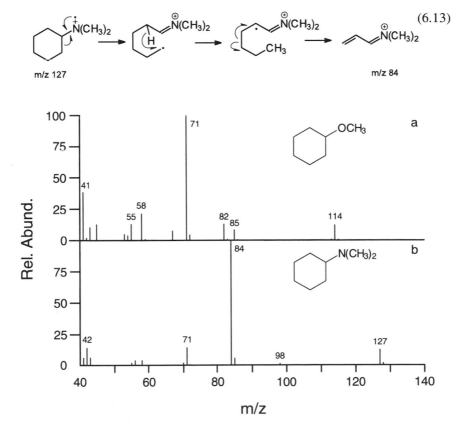

**Figure 6.12.** Generalized mechanism for the cyclohexanone-type rearrangement. X=OR, NR$_2$, phenyl, =O.

**Figure 6.13.** Mass spectra of two cyclohexane derivatives in which ions due to the cyclohexanone-type rearrangement are the dominant feature: (*a*) methoxycyclohexane and (*b*) *N,N*-dimethylaminocyclohexane.

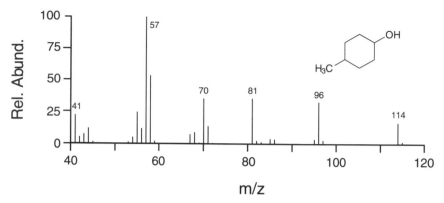

**Figure 6.14.** Mass spectrum of 4-methylcyclohexanol (Problem 6.2).

The abundance of these peaks relative to other ions in the spectra, especially in the case of the amine, reflects the greater ability of aliphatic oxygen and nitrogen to stabilize the resultant positive charge, as opposed to the carbonyl oxygen.

***Problem 6.2.*** The mass spectrum of 4-methylcyclohexanol is shown in Figure 6.14. Write a mechanism to account for the formation of the base peak at *m/z* 57.

***Problem 6.3.*** Phencyclidine [1-(1-phenylcyclohexyl)piperidine; PCP or angel dust] is a veterinary tranquilizer that has potent (and dangerous) hallucinogenic properties. This compound, whose structure is shown below, has a cyclohexane ring with two functional groups able to direct α-cleavage. Account for formation of the base peak at *m/z* 200 in the spectrum of this compound.

## 6.4. RETRO DIELS–ALDER REARRANGEMENT

In γ-hydrogen and cyclohexanone-type rearrangements, six-membered cyclic transition states are formed. On the other hand, compounds that already have six-membered rings containing one double bond (cyclohexene derivatives) may undergo a fragmentation in which the ring is cleaved to produce an olefin and a diene. The approximate reverse of this reaction in neutral molecules—the Diels–Alder reaction (named after the Nobel-Prize-winning chemists who discovered it) is important in synthetic organic chemistry for constructing six-membered ring systems:

$$\text{heat} \qquad (6.14)$$

Many substituents are possible in the retro Diels–Alder fragmentation, which proceeds even in compounds having heteroatoms and complex ring structures. The energy factors that control this reaction are sensitive to subtle structural changes, so that the retro Diels–Alder fragmentation may produce a large peak in the spectrum of one compound, but a closely related compound may show little or no evidence for this fragmentation at all.

The retro Diels–Alder fragmentation may proceed via two different mechanisms (Figure 6.15). In both cases initial ionization occurs at the double bond, followed by electron density redistribution in which the positive charge either is retained on the incipient diene fragment (charge retention) or transferred to the olefin product (charge migration). It is tempting to write out this fragmentation using double-headed arrows as a shorthand notation:

$$\qquad (6.15)$$

While the Diels–Alder reaction often proceeds in a concerted manner without the formation of intermediates, the retro Diels–Alder fragmentation does not, and understanding the outcome of this fragmentation is more difficult when both mechanisms and all possible intermediates are not considered.

Limonene [(1-methyl-4-(2-propenyl)cyclohexene)], a terpene largely responsible for the characteristic odor of lemons, contains a cyclohexene ring having only alkyl substituents. The retro Diels–Alder fragmentation accounts for the base peak in its mass spectrum [Figure 6.16 and Eq. (6.16)]:

Charge-retention:

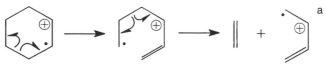

Charge-migration:

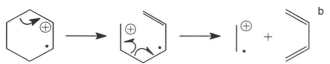

**Figure 6.15.** The retro Diels–Alder fragmentation can proceed either by charge retention or charge migration. (*a*) Charge-retention (charge retained by the diene fragment); (*b*) charge-migration (charge transferred to olefin fragment). Many substituents are possible.

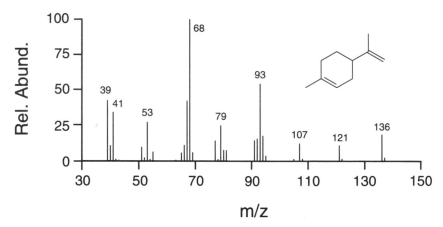

**Figure 6.16.** Mass spectrum of limonene. The retro Diels–Alder fragmentation accounts for the base peak.

(6.16)

m/z 136                                    m/z 68

Because the molecule splits in half during this reaction, both the charge retention and charge migration mechanisms predict the same products.

The mass spectra of 3- and 4-phenylcyclohexene, on the other hand, show different responses to the retro Diels–Alder reaction (Figure 6.17). If mechanisms for these fragmentations are written only in shorthand form, we would predict that the 4-phenyl isomer should produce either butadiene ($m/z$ 56) or styrene molecular ion ($m/z$ 104):

(6.17)

m/z 158          m/z 104                   or           m/z 56

while 3-phenylcyclohexene should produce either ethylene ($m/z$ 28) or 1-phenylbutadiene molecular ion ($m/z$ 130):

(6.18)

m/z 158          m/z 28              or                 m/z 130

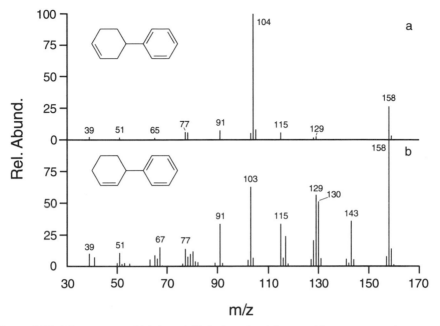

**Figure 6.17.** Mass spectra of (*a*) 4- and (*b*) 3-phenylcyclohexene. These compounds respond in different ways to the retro Diels–Alder fragmentation.

On the basis of product stability alone, it would seem that, because of its extended conjugation, the *m/z* 130 ion in the 3-phenylcyclohexene spectrum should form most readily. It is thus puzzling that the *m/z* 104 ion dwarfs almost all the other ions in the spectrum of the 4-phenyl isomer, while the *m/z* 130 ion in the spectrum of the 3-phenyl isomer is only one of many important fragment ions in the spectrum.

A closer examination of the fragmentation mechanisms helps clear up this confusion. In the case of 4-phenylcyclohexane, *m/z* 104 forms by charge migration since the positive charge in the product ion must be transferred from the double bond (the site of original ionization) to the other side of the cyclohexane ring. A charge retention mechanism is needed to account for the formation of *m/z* 130 in the 3-phenyl isomer. Theoretically, ionization of the double bond can occur so that the positive charge is located on either carbon in both of these compounds. Initial ionization as depicted in Eqs. (6.19) and (6.20) facilitates writing these mechanisms:

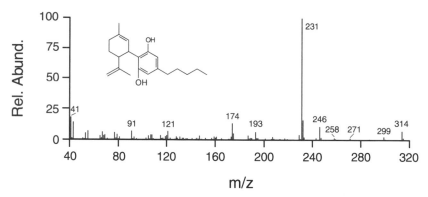

**Figure 6.18.** Mass spectrum of cannabidiol. The ion at $m/z$ 246 results from retro Diels–Alder fragmentation.

In Eq. (6.19) the first step of the rearrangement involves neutralization of the originally formed positive charge by transfer of an electron pair and relocation of the charge on the benzylic carbon. The other product formed in this step is an allylic radical, which also is resonance-stabilized. On the other hand, the first step in the fragmentation of 3-phenylcyclohexene in Eq. (6.20) produces an ion that is resonance stabilized by extended conjugation, but it also produces a primary radical. Evidently the small amount of additional stabilization energy gained by extending the conjugation does not adequately compensate for the difference in energy between an allylic and a primary radical, so that formation of $m/z$ 104 in the spectrum of 4-phenylcyclohexene requires less energy than formation of $m/z$ 130 in the 3-phenyl isomer.

Although not often the primary mode of fragmentation, the retro Diels–Alder fragmentation also occurs in more complex molecules. The spectrum of cannabidiol, the biosynthetic precursor to $\Delta^9$-THC in marijuana, exhibits few intense fragment ions at high mass (Figure 6.18). As with limonene above (note the similarities in structure!), the ion at $m/z$ 246 results from the loss of 68 daltons from the molecular ion by charge retention retro Diels–Alder rearrangement as shown in Eq. (6.21):

(6.21)

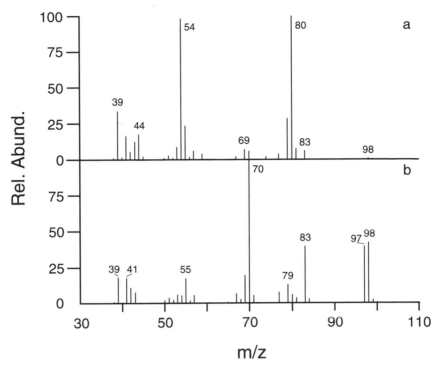

**Figure 6.19.** Mass spectra for Problem 6.4.

[Eq. (6.21) adapted with permission from Smith, 1997. Copyright ASTM.] This is a rare instance where initial ionization at an isolated double bond must be invoked, despite the presence of heteroatoms in the molecule, in order to account for the observed products. The ion thus formed gains further stability by cyclizing to form a second six-membered ring, followed by a four-center hydrogen migration to place the radical ion site on the oxygen. α-Cleavage of one of the adjacent methyl groups [as in THC and cannabinol; Figure 6.7 and Eq. (5.11), respectively] produces the highly stable aromatic ion at $m/z$ 231 (Smith, 1997).

**Problem 6.4.**  The spectra of 3- and 4-hydroxycyclohexene are shown in Figure 6.19. As you can see, the spectra are very different. Which spectrum goes with which isomer? Give a reason for your answer.

3-Hydroxycyclohexene          4-Hydroxycyclohexene

***Problem 6.5.*** $\Delta^8$-THC (MW 314) has a significant peak in its mass spectrum at $m/z$ 246, but $\Delta^9$-THC does not. Explain this difference in behavior.

$\Delta^8$-THC                                   $\Delta^9$-THC

## 6.5. DOUBLE-HYDROGEN REARRANGEMENT

Not all mass spectral rearrangements proceed via three-, four-, and six-center transition states. In the spectrum of butyl palmitate (Figure 6.20), the ions resulting from McLafferty rearrangement [$m/z$ 116 and 256; Eq. (6.3)] account for only a small proportion of the total fragmentation. Instead, the major fragment ion at high mass occurs at $m/z$ 257, one dalton higher than the ion from McLafferty rearrangement on the ester side of the carbonyl group.

The fragmentation producing this ion, which is characteristic of esters and amides derived from alcohols and amines having propyl aliphatic chains or larger, involves the transfer of *two* hydrogens and thus is known as the double-hydrogen rearrangement. The migration of the second hydrogen is somewhat unusual in that it involves formation of a five-membered cyclic transition state:

(6.22)

The adverse entropy contribution that comes from forming the five-membered ring is overcome by the fact that both the ion and radical products are resonance-stabilized.

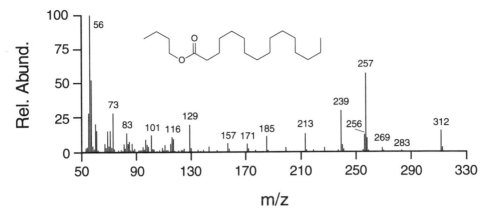

**Figure 6.20.** Mass spectrum of butyl palmitate. The ions at *m/z* 56 and 257 result from the double-hydrogen rearrangement.

## REFERENCE

R.M. Smith, *J. Forensic Sci.,* **42,** 608–616 (1997).

# CHAPTER 7

# WRITING MASS SPECTRAL FRAGMENTATION MECHANISMS

## 7.1. GENERAL GUIDELINES

Many spectra contain peaks from unexpected fragmentations—ones that do not seem to fall obviously into any previously encountered category. These must be rationalized "after the fact," usually without help from independent mechanistic studies such as deuterium labeling. Writing mechanisms for simple and expected mass spectral fragmentations may be straightforward; many examples have been presented so far in this book. Fragmentations of complex molecules, however, are sometimes so involved that they demand insight and ingenuity to rationalize. It is possible, nonetheless, to set down some general guidelines for devising fragmentation mechanisms. Most of these priniciples have been introduced in previous chapters, but it is worthwhile to summarize them here. In the remainder of this chapter, we will explore the application of these guidelines to two specific examples.

1. Only *intra*molecular reactions are allowed (Chapter 1). The only exceptions to this are the formation of reagent gas and protonated molecules $(M + H)^+$ in chemical ionization mass spectrometry.

2. Compounds with heteroatoms ionize initially by loss of an electron from the *n*-orbital(s) of the heteroatom(s). For compounds having double bonds or aromatic rings, initial loss of one of the $\pi$-electrons is *possible* if heteroatoms are present and is preferred if heteroatoms are absent. Initial ionization by loss of $\sigma$-electrons is a last resort and, for our purposes, occurs only in aliphatic hydrocarbons (Chapter 3).

3. Charges and electrons must be balanced for each step of a fragmentation sequence (Chapter 3). Losing, or gaining, an electron is an easy way to get off the correct reaction path.

4. The mass of each intermediate must be verified; errors in simple arithmetic probably account for more mistakes in mass spectral interpretation than any other single factor. After verifying the mass of the intermediate, the size of the lost fragment can be calculated and possible structures explored (e.g., using Table 4.1). The structure of the parent ion is then examined for probable origins of the lost fragment.

5. As a general rule, mechanisms for forming major ions (especially high mass ions) involve few steps. Exceptions may occur if the ring structure of the molecule is so complex that simple fragmentations lead only to isomeric ionic structures, not to actual fragmentation (see Chapter 8). Mechanisms that require many intermediates usually have greater energy demands than those that do not, and are thus less likely to occur (Chapter 3).

6. The most labile bonds in the molecule (those that are electron-poor) break most easily. α-Cleavage (Chapter 5) is an excellent example of this type of fragmentation, occuring in virtually every molecule in which it can take place. It is important not to overlook the possibility for α-cleavage "by long-distance," that is, by initial ionization at a heteroatom, followed by cleavage at a site several bonds removed from the heteroatom and subsequent reorganization of the electron density to stabilize the positive charge on the heteroatom. For this to occur the heteroatom and the cleavage site must be connected by one or more double bonds [see, e.g., Eq. (5.12)].

7. The fragmentations and rearrangements discussed in this book, because of their simplicity and general applicability, should be considered as starting points and patterns for mechanism writing. Two prominent categories include

    a. Loss of small, unsaturated molecules with concurrent formation of an ion that is at least as stable as the parent ion (e.g., Section 4.3)

    b. Fragmentations in which the positive charge is stabilized either by conjugation with a double bond or aromatic system or by a heteroatom (Chapter 5)

8. Hydrogens are easily movable. The energy demands of rearranging a hydrogen radical from one site to another is often so low that short-term decreases in ion stability are tolerated if the overall sequence forms stable ion and radical products. Hydrogen migration occurs most often via three-, four-, and six-membered cyclic transition states, with five- and seven-membered rings occurring less frequently due to entropy demands on the energy of activation (Chapter 3).

9. When one resonance structure does not give the desired result, another might work. This can include ring openings and closings that have parallels in solution chemistry, such as:

    a. Benzyl ↔ tropylium:

$$(7.1)$$

b. Unsaturated seven-membered rings ↔ bicyclo[4.1.0]heptyl systems:

(7.2)

c. Unsaturated eight-membered rings ↔ bicyclo[4.2.0]octyl systems:

(7.3)

## 7.2. EPHEDRINE

A molecule complex enough to be challenging, yet whose spectrum is amenable to rationalization, is ephedrine, a mild stimulant and decongestant. The mass spectrum of ephedrine (Figure 7.1a) is similar to that of methamphetamine (Figure 7.1c), showing fragmentation dominated by $\alpha$-cleavage to give the intense ion at $m/z$ 58:

(7.4)

The obvious similarities end at that point, however. Ephedrine shows no molecular ion at $m/z$ 165, and the first observed loss (19 daltons!!) produces the small ion at $m/z$ 146. The other ion resulting from $\alpha$-cleavage ($m/z$ 107) is somewhat overshadowed by the ions at $m/z$ 106 and 105 as well as by those at $m/z$ 77 to 79. It is especially important to rationalize the formation of these ions since the spectrum of methcathinone (Figure 7.1b), another powerful stimulant that appeared recently on the illicit drug scene, is even more similar to that of ephedrine than is that of methamphetamine.

Since ephedrine contains no fluorine, the loss of 19 daltons cannot occur in one step. Two combinations are worth considering—one involving the loss of $NH_2 \cdot$ or $NH_3$ and either three or two hydrogens, and the second involving loss of water and hydrogen radical. Although methamphetamine loses $CH_3NH$ radical to produce the small ions at $m/z$ 115 to 119 [Eq. (7.5)], the loss of $NH_2 \cdot$ or $NH_3$ is not observed. There is no reason to assume that ephedrine would lose either of these fragments.

On the other hand, loss of water is well documented from some alcohols, involving prior rearrangement of hydrogen from a site well removed from the oxygen (Section 5.4.2). In ephedrine, following initial ionization at oxygen, transfer of a hydro-

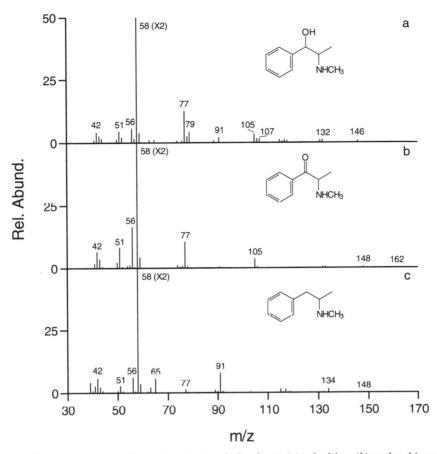

**Figure 7.1.** Mass spectra of three closely related stimulants: (*a*) ephedrine, (*b*) methcathinone, and (c) methamphetamine.

(7.5)

gen from the *N*-methyl group to the oxygen can occur via a six-membered cyclic transition state:

(7.6)

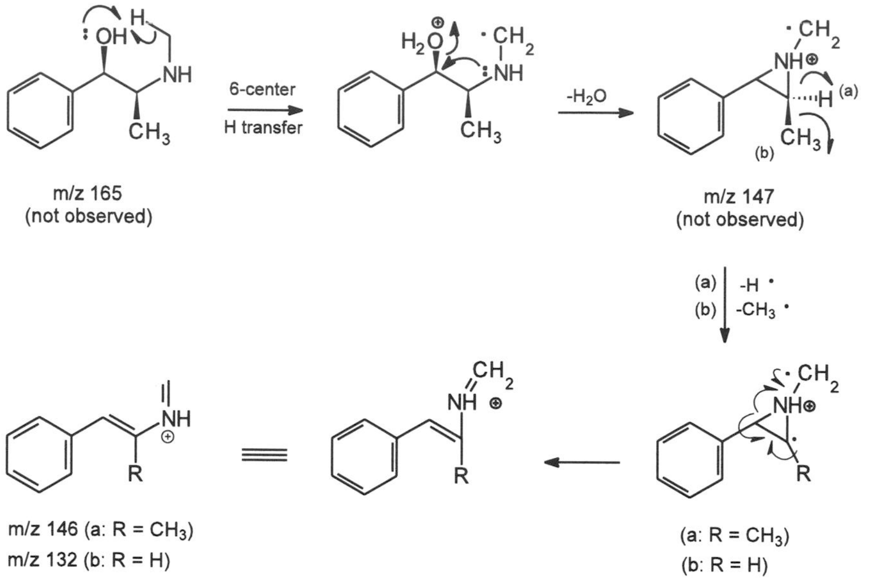

m/z 165
(not observed)

6-center
H transfer

$-H_2O$

(a)

m/z 147
(not observed)

(a) | -H ·
(b) | -CH₃ ·

m/z 146 (a: R = CH₃)
m/z 132 (b: R = H)

(a: R = CH₃)
(b: R = H)

Nucleophilic displacement of water by the lone pair of electrons on nitrogen then places the positive charge on nitrogen (*m/z* 147). This radical ion is not observed since the primary radical site and highly strained three-membered ring create enough instability that the activation energy for further fragmentation is lowered—apparently enough that it fragments before leaving the ion source.

The *m/z* 147 ion can relieve these instabilities through loss of either hydrogen or methyl radical and concurrent rearrangement of the electron density in the ion [Eq. (7.6)]. Both products of this fragmentation are stabilized by extended conjugation with the aromatic ring, which explains why phenyl radical, which is similarly situated with respect to the nitrogen, is not lost instead. It also explains why the direct loss of either a methyl or a hydrogen radical by α-cleavage is not observed. The relative intensities of the *m/z* 146 and 132 ions are consistent with the greater stability of methyl over hydrogen radical when the ion products have similar stabilities [compare Eq. (5.2)].

Although initial ionization at nitrogen leads to α-cleavage with loss of the hydroxybenzyl radical and formation of the ion at *m/z* 58 [Eq. (7.4)], formation of hydroxybenzyl ion (*m/z* 107) does not require a charge migration mechanism:

(7.7)

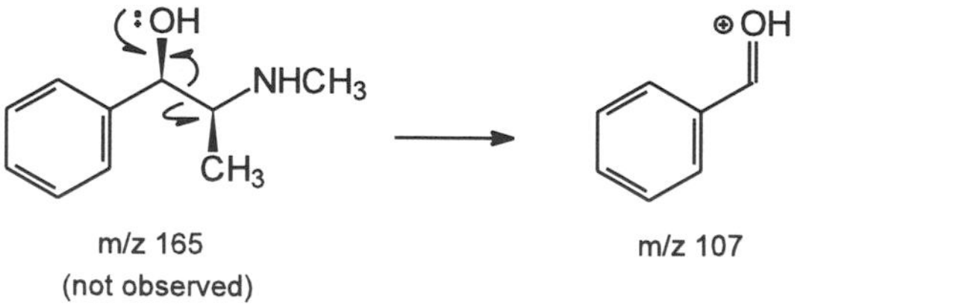

m/z 165
(not observed)

m/z 107

[compare Eq. (5.3)]. Further, this ion is unstable toward loss of hydrogen to give the benzaldehyde radical ion (*m/z* 106) and the benzoyl ion (*m/z* 105; Figure 7.2). Although the phenyl ion (*m/z* 77) is expected by loss of CO from the benzoyl ion, the ions at *m/z* 78 and 79 are more difficult to explain since the spectrum of benzaldehyde (Figure 5.20b) contains neither of them.

The ion at *m/z* 79 can arise by loss of CO from the *m/z* 107 ion, although rearrangement of hydrogen from both the oxygen and the adjacent carbon must occur prior to this loss (loss of ethylene from *m/z* 107 seems unlikely since aromatic ions prefer to lose acetylene instead; Figure 4.7). Two rearrangements that have parallels in solution chemistry help account for this loss. First, the *m/z* 107 ion has two other forms that are more useful. One has the positive charge located on the benzylic carbon (the true hydroxybenzyl ion), while the second, because of the equilibrium that exists between benzylic ions and their cycloheptatrienyl counterparts [Eq. (7.1)], is the hydroxytropylium ion (Figure 7.2). The hydroxytropylium ion, containing an "aromatic" hydroxyl group, is similar to phenol and, as such, is in equilibrium with its keto form by means of a four-center hydrogen transfer [compare Eq. [4.4)]. These equilibria achieve the goal of transferring the hydrogens from the oxygen and its adjacent carbon to other parts of the molecule. Even though the keto form is much less stable than the enol form due to the loss of aromatic resonance energy, this equilibrium will constantly be disturbed by the irreversible formation of other ion products.

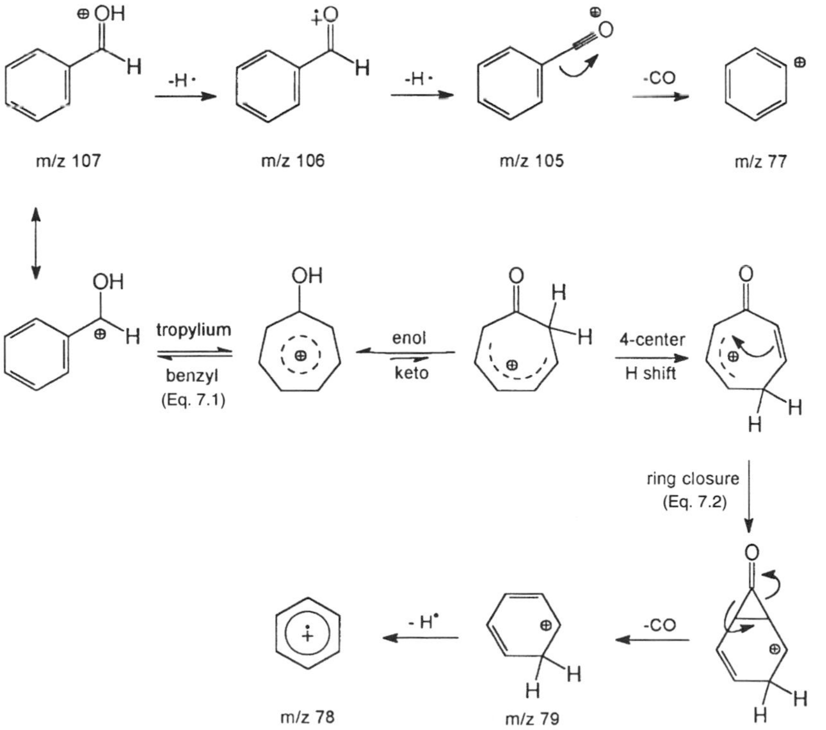

**Figure 7.2.** Fragmentations of the hydroxybenzyl ion (*m/z* 107) from ephedrine.

Loss of carbon monoxide may proceed directly from the keto form, although the following rationalization, which also has parallels in solution chemistry, seems easier to justify. Four-center transfer of hydrogen, followed by closure of the unsaturated seven-membered ring to form a bicyclo[4.1.0]heptenyl ion [Eq. (7.2)], isolates the carbon-monoxide-producing moiety so that cleavage of the strained cyclopropanone ring yields the desired result. The ion at $m/z$ 79 is unstable toward hydrogen radical loss to give the benzene molecular ion ($m/z$ 78). Further loss of hydrogen to give the phenyl ion seems less likely since benzene itself does this inefficiently.

***Problem 7.1.*** Write mechanisms for the formation of the ions at $m/z$ 148, 105, 77, 58, and 51 in the spectrum of methcathinone (Figure 7.1*b*). Why don't the same fragmentations that occur in ephedrine take place in this compound? Are the mass spectra of these two compounds different enough to permit their identification by mass spectrometry?

## 7.3. *ORTHO* EFFECT—THE ETHOXYBENZAMIDES

Considering the similarity in their structures, the mass spectra of *o*- and *m*-ethoxybenzamide could hardly be more different (Figure 7.3). The high-mass ends of the spectra have few peaks in common, and even those ions found in both spectra differ

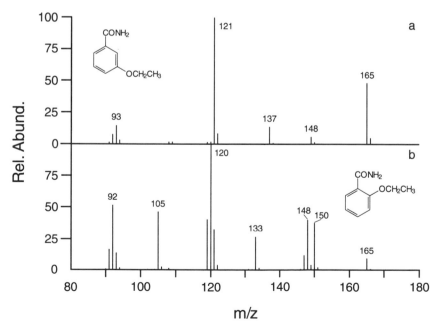

**Figure 7.3.** Mass spectra of (*a*) *m*-ethoxybenzamide and (*b*) *o*-ethoxybenzamide are very different, due to interactions that are possible between the substituents in the *ortho-is*omer.

substantially in relative abundance. Clearly, rationalizing the difference in behavior between these two compounds is desirable.

The key to this difference is found in the spectrum of the *para*-isomer, which is similar to that of the *meta*-isomer. Apparently the proximity of the two functional groups in the *ortho*-compound allows them to interact in ways that are not possible in the other two isomers. This "*ortho* effect" frequently occurs in multiply substituted benzene derivatives (Section 6.2.3).

The spectrum of *m*-methoxybenzamide is readily explainable in terms of previously discussed fragmentations (Figure 7.4). After initial ionization at the carbonyl oxygen, the molecular ion loses ·NH$_2$ by α-cleavage typical of primary amides to give the *m/z* 149 ion. Formation of the ion at *m/z* 121 can occur by several pathways, the most logical being by loss of CO from *m/z* 149.

On the other hand, initial ionization at the ethoxy oxygen, followed by a four-centered migration of hydrogen and loss of ethylene [see Eq. (5.21)], leads to the ion at *m/z* 137. Formation of the ion at *m/z* 93 can occur from either *m/z* 137 by α-cleavage of the amide group and elimination of CO, or by ethylene loss from *m/z* 121.

In contrast to the facile loss of the amide group shown by the *meta*-isomer, *o*-ethoxybenzamide forms the following ions via pathways that must be energetically more favorable due to interactions between the two functional groups:

(a)  *m/z* 150 (loss of methyl radical)

(b)  *m/z* 148 (loss of 17 daltons, with virtually no loss of 16 daltons)

(c)  *m/z* 133

(d)  *m/z* 120 (loss of 45 daltons)

(e)  *m/z* 105

(f)  *m/z* 92

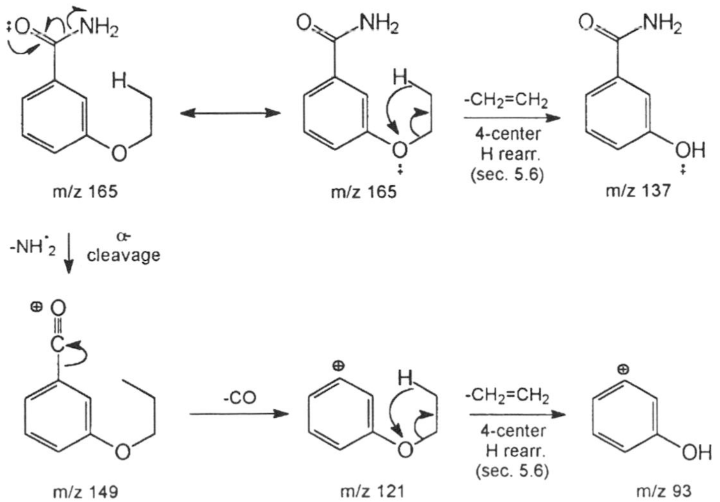

**Figure 7.4.** High mass fragmentations of *m*-methoxybenzamide.

*m/z 150.* Since the *meta*-isomer does not lose methyl, it seems likely that this fragmentation does not involve simple α-cleavage after initial ionization at the ethoxy oxygen. On the other hand, the intensity of this ion implies the straightforward loss of methyl from the ethoxy group rather than loss of a different carbon accompanied by rearrangement of hydrogen. Initial ionization at nitrogen, followed by loss of methyl radical with formation of a six-membered ring and stabilization of the positive charge on the nitrogen, conveniently rationalizes this fragmentation using the guidelines in Section 7.1:

(7.8)

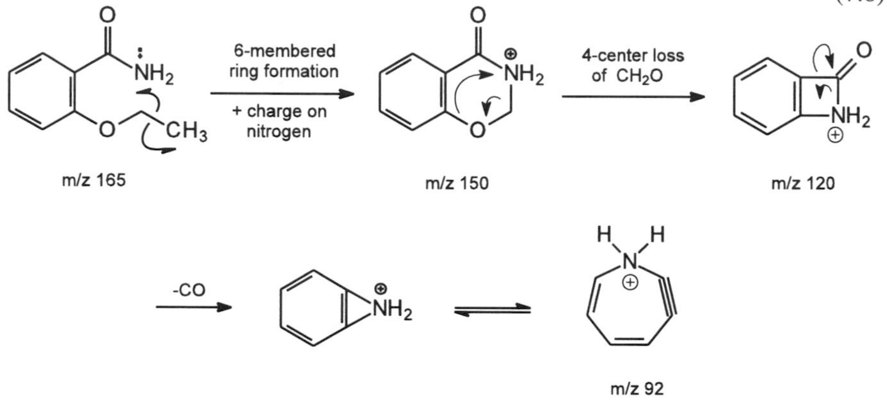

*m/z 148.* Although loss of 17 daltons as hydroxy (Table 4.1) cannot be ruled out without additional work, this fragmentation would appear to require more energy than those described below. One explanation that seems appealing is loss of $H_2$ from *m/z* 150 to form an ion having extended conjugation:

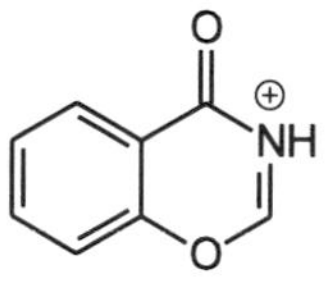

However, the *m/z* 149 has virtually no intensity beyond the isotopic contributions from *m/z* 148, and hydrogen loss such as this usually occurs in a stepwise fashion, showing at least some evidence of the ion in which only one hydrogen is lost.

   An alternative explanation is loss of ammonia. Initial ionization at the ethoxy oxygen, hydrogen radical loss by α-cleavage (not observed at *m/z* 164, but these ions often are very weak; Chapter 5), and elimination of $NH_2 \cdot$ with ring closure leads to the molecular ion of an aromatic compound (Figure 7.5).

*m/z 147 and 133.* The structure proposed for *m/z* 148 is substantiated by the relative intensities of the *m/z* 147 and 133 ions, whose structures have extended conju-

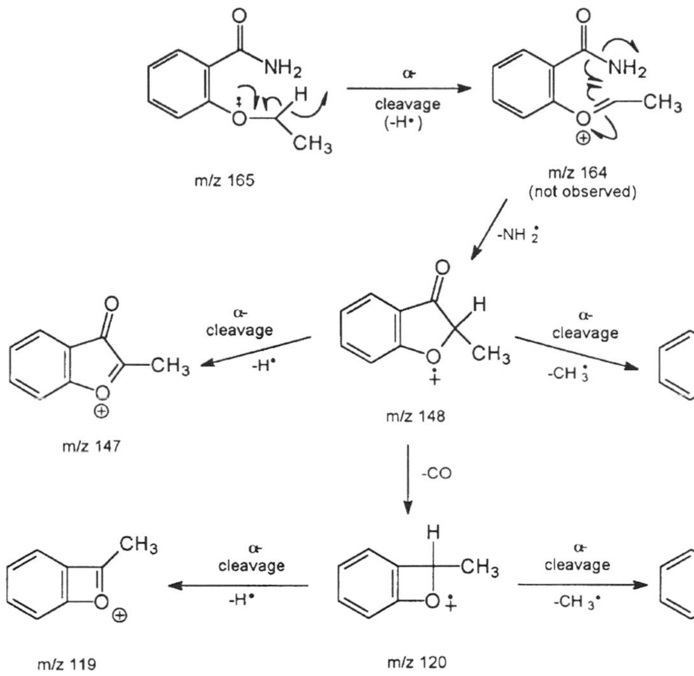

**Figure 7.5.** Formation of several ions in *o*-ethoxybenzamide through a common intermediate at *m/z* 148.

gation and can be formed via α-cleavage loss of hydrogen and methyl radicals, respectively (Figure 7.5). Formation of *m/z* 133 by loss of ammonia from *m/z* 150 seems less likely since this would entail reopening of the ring and further hydrogen rearrangement.

*m/z 120.* The structure of the ion at *m/z* 120 is problematic; in fact, two different structures for this ion can be supported by the data, and both may well occur. High-resolution mass spectrometry would resolve this issue. Loss of CO from *m/z* 148 can produce the molecular ion of an aromatic compound whose stability is approximately the same as that of the parent ion. (Loss of ethylene from this ion cannot proceed without substantial rearrangement). This *m/z* 120 ion, like *m/z* 148, can lose either hydrogen or methyl radical by α-cleavage to produce *m/z* 119 and 105 (Figure 7.5). Notice that the *m/z* 105 ion thus formed is *not* the benzoyl ion but is isomeric with it. Support for this structure comes from the "benzoyl" low mass ion series in which the sizes of the *m/z* 51 and 77 ions are much smaller relative to *m/z* 105 than those normally encountered from benzoyl (Section 5.4.3). This indicates that the energy necessary to form *m/z* 77 and 51 from the proposed structure is greater than from benzoyl, consistent with the fact that this ion must rearrange hydrogen onto the aromatic ring before eliminating CO.

The other structure for $m/z$ 120 arises from formal loss of ethoxy radical. Initial ionization at the nitrogen and formation of a small ring in which the positive charge is stabilized by nitrogen is possible:

$$(7.9)$$

m/z 165                              m/z 120

However, an alternative mechanism involving loss of formaldehyde from $m/z$ 150 is somewhat easier to justify since the parent and fragment ions in this case have similar stabilities and fragmentation is driven by loss of a small, unsaturated molecule [Eq. (7.8)], rather than by radical formation and the severing of a carbon–heteroatom bond. In either case, support for this structure is found in the size of the $m/z$ 92 ion, which can form by subsequent loss of CO.

Although other mechanisms can be proposed, formation of all of the major high mass ions in this spectrum has been explained by invoking only two intermediates— the ions at $m/z$ 150 and 148. The formation of $m/z$ 146 and 132 in the spectrum of ephedrine (Section 7.2) by a single intermediate is equally appealing.

**Problem 7.2.** The spectra of methyl 3,5-dimethylbenzoate and methyl 2,5-dimethyl-benzoate (Figure 7.6) both show molecular ions losing methoxy radical by α-cleavage to give a base peak ion at $m/z$ 133, which subsequently loses CO to give an intense ion at $m/z$ 105. However, the latter compound also forms intense ions at $m/z$ 132 (apparently from loss of methanol) and $m/z$ 104 that are not present in the spectrum of the 3,5-isomer. Rationalize this difference.

**Problem 7.3.** Formation of the $m/z$ 271 ion in the spectrum of $\Delta^9$-tetrahydro-cannabinol (Figure 6.6a) is puzzling. This ion appears to arise not from the loss of $CH_2{=}CH_2$ or CO from $m/z$ 299, but rather from loss of $C_3H_7$ from the molecular ion, including loss of the *gem*-dimethyl group plus another hydrogen. Write a mechanism for this fragmentation. (*Hint:* List all possible products from α-cleavage of the molecular ion. Do any of these look promising for rearranging a hydrogen to the carbon having the *gem*-dimethyl group?)

**Problem 7.4.** The mass spectrum of the indole alkaloid harmine is shown in Figure 7.7. Write reasonable mechanisms for formation of the two fragment ions at $m/z$ 197 and 169. (*Hint:* Which methyl group is most likely to be lost?)

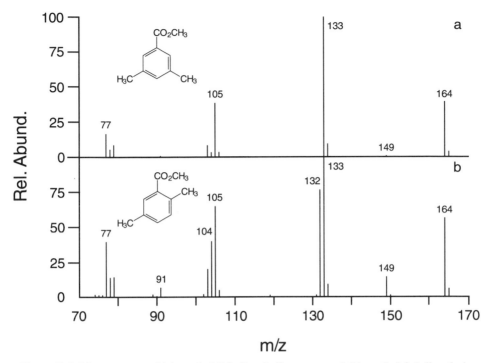

**Figure 7.6.** Mass spectra of (*a*) methyl 3,5-dimethylbenzoate and (*b*) methyl 2,5-dimethylbenzoate (Problem 7.2).

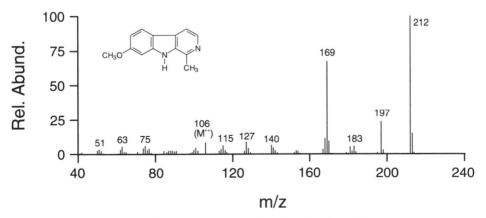

**Figure 7.7.** Mass spectrum of harmine (Problem 7.4).

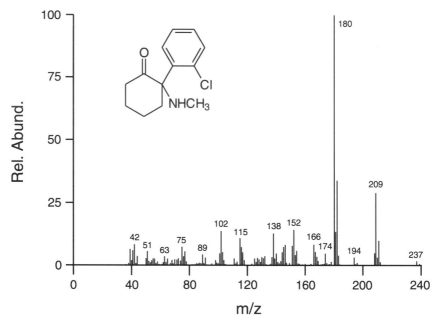

**Figure 7.8.** Mass spectrum of ketamine (Problem 7.5).

***Problem 7.5.*** Ketamine, whose structure and mass spectrum are shown in Figure 7.8, is a veterinary tranquilizer (like phencyclidine) that also produces hallucinogenic effects. Account for formation of the ions at *m/z* 209, 194, and 180 in this spectrum.

# CHAPTER 8

# STRUCTURE DETERMINATION IN COMPLEX MOLECULES USING MASS SPECTROMETRY

## 8.1. INTRODUCTION

The mass spectra of large molecules are often so complex that they literally defy complete interpretation. In such cases, the most we can usually hope to do is to rationalize the formation of most of the high mass ions and other ions of significant intensity. Even this is not possible in all cases; the mass spectra of some steroids, for example, have so many intense ions that these molecules seem to break apart just about everywhere (Figure 8.1)! Unfortunately (from an interpretative standpoint) most of the molecules encountered in day-to-day organic analysis are complex. Chemists analyzing drugs of abuse, pesticides, natural products, or the products of organic syntheses routinely work with molecules having molecular weights of up to several hundred daltons. Biomolecules, some with molecular weights in the thousands of daltons, are no longer unapproachable, given the development of specialized methods for sample introduction, ionization, and volatilization (see Table 1.1).

Although the fragmentations discussed in this book are applicable to large molecules, as should be apparent from the examples given so far, we need a somewhat different approach to structure determination when dealing with the mass spectra of these compounds. The reasons are simple. First, it is nearly impossible to determine a unique structure for an unknown complex molecule solely on the basis of its mass spectrum, unless the structure of the compound has been determined previously by independent means and a "standard" mass spectrum already exists for the compound. Second, the interaction of the various functional groups present in these compounds often leads to fragmentation pathways that are difficult to predict and sometimes hard even to rationalize.

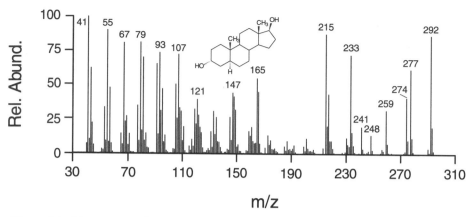

**Figure 8.1.** Mass spectra of some complex molecules, such as that of the steroid shown here, may contain so many intense ions that only the formation of a few high mass ions can be rationalized.

Many approaches are possible for interpreting the spectra of large molecules. Some of the tools that help identify specific fragmentation processes—preparation of deuterated derivatives, high resolution MS, and MS/MS (see Section 1.3.4)—are either extremely time-consuming or demand instrumentation beyond the resources or goals of small analytical labs. A more practical approach is to study the spectra of families of compounds that possess the same basic structure but that differ in the functional groups attached to them. This is similar to studying the spectra of deuterated derivatives, except that, in some cases, the work of characterizing the derivatives may already have been done—either in our own laboratories or by others working within the same narrow area of organic analysis. A disadvantage of this approach is that, because other functional groups are involved, the derivatives may fragment in ways that differ substantially from those of the parent compound.

In this chapter the mass spectra of four families of drugs of abuse will be discussed. In each instance the spectra of different family members will be compared in order to understand as much as possible how the complex structures of these molecules fragment. This information then will be used in a predictive manner to determine unique structures for other members of the family solely by means of their mass spectra. In the process, we will encounter further examples of the fragmentations and techniques of mechanism writing discussed in previous chapters.

## 8.2. "DESIGNER DRUGS" RELATED TO MDA

3,4-Methylenedioxyamphetamine (commonly known as MDA) is a synthetic hallucinogen that has some structural similarities to mescaline, the active drug found in the peyote cactus. The mass spectrum of MDA (Figure 8.2a) is similar to that of amphetamine (Figure 5.9). Prominent low mass aromatic ions occur at *m/z* 51 and 77,

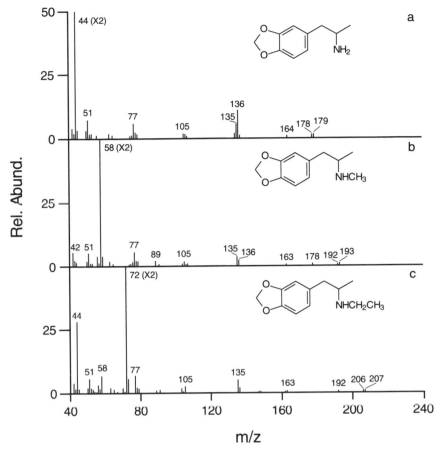

**Figure 8.2.** Mass spectra of (*a*) 3,4-methylenedioxyamphetamine (MDA) and two of its *N*-alkylated derivatives: (*b*) *N*-methyl MDA (MMDA), and (*c*) *N*-ethyl MDA (MDEA).

and charge retention α-cleavage accounts for the base peak [with loss of the benzylic radical; Eq. (8.1)]:

$$\text{(8.1)}$$

m/z 44 (R = H)
m/z 58 (R = Me)
m/z 72 (R = Et)

as well as the ions at *m/z* 178 and 164 (with losses of hydrogen and methyl radicals, respectively). In contrast to amphetamine, formation of *m/z* 135 can be rationalized by "long-distance" α-cleavage with charge retention by the benzylic fragment:

(8.2)

m/z 135

The ion at $m/z$ 136 is reminiscent of γ-hydrogen rearrangement ions found in the spectra of alkylbenzenes (Section 6.2.2), but in this case the migrating hydrogen probably does not come from the terminal methyl group since other propylbenzenes having heteroatoms on the ring do not undergo this rearrangement (see Figure 6.7). Migration of one of the amine hydrogens seems more likely since the intensity of $m/z$ 136 for the N-alkyl derivatives (see below) decreases relative to that for MDA (Figure 8.2). Since amphetamine and methamphetamine show a much diminished propensity to undergo this fragmentation (compare the relative abundances of $m/z$ 77 and 78 in Figures 5.9 and 5.10), these data indicate that the methylenedioxyphenyl ring provides an electron-rich site for attracting the migrating hydrogen:

(8.3)

m/z 136

Illicit drug manufacturers have repeatedly attempted to thwart prosecution by synthesizing new compounds (called designer drugs) that have the same basic chemical structures as controlled drugs but that do not fall specifically under existing statutes (and thus are legal to possess and sell—at least until legislation is passed that covers them) because they have slightly different functional groups attached to them. Two hallucinogens that fall into this category (both are now controlled drugs!) are 3,4-methylenedioxymethamphtamine (MDMA; "Ecstacy") and 3,4-methylenedioxy-N-ethylamphetamine (MDEA; "Eve"), N-alkyl homologs of MDA.

The most obvious difference between the mass spectra of these two compounds and that of MDA (Figure 8.2) is the position of the base peak. This shift is predictable, and is consistent with primary α-cleavage loss of the methylenedioxybenzyl radical and charge stabilization on the nitrogen [Eq. (8.1)]. Both compounds also show minor losses of hydrogen and methyl by α-cleavage and the formation of $m/z$ 135 by benzylic cleavage. The spectrum of MDEA also exhibits an intense $m/z$ 44 ion, re-

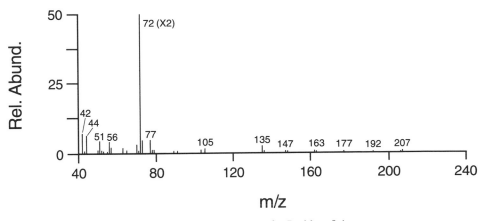

**Figure 8.3.** Mass spectrum for Problem 8.1.

sulting from loss of ethylene via secondary rearrangement from the initially formed α-cleavage ion:

$$(8.4)$$

(see Section 5.5). The small ions at $m/z$ 163 in the spectra of MDMA and MDE are undoubtedly due to loss of the *N*-alkylamino radical, similar to the behavior of amphetamine and methamphetamine [Eq. (7.5)].

*Problem 8.1.* A compound suspected of being a new MDA analog gave the mass spectrum in Figure 8.3. Can you assign a unique structure to this compound based on this spectrum?

## 8.3. COCAINE AND ITS METABOLITES[1]

The mass spectrum of MDA is dominated by two major types of fragmentation—α-cleavage and the γ-hydrogen rearrangement. This makes interpreting the spectrum straightforward and predicting the behavior of previously unencountered analogs relatively easy. Interpreting the mass spectrum of the more complex structure of cocaine (Figure 8.4), on the other hand, is more difficult.

Determining how cocaine fragments is facilitated by the substantial literature on the metabolism of this drug (Figure 8.5), including the identification of many new metabolites by mass spectrometry. Compiling and analyzing the data in these studies

[1]A more thorough discussion of the material in this section has been published in R.M. Smith (1997).

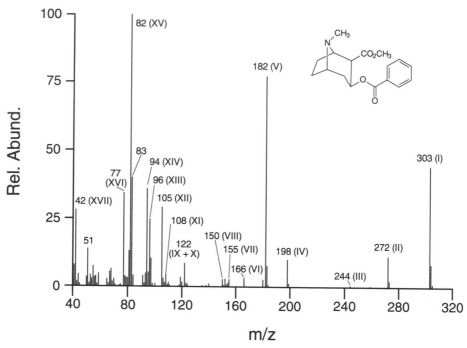

**Figure 8.4.** Mass spectrum of cocaine. Roman numerals refer to ion structures listed in Table 8.1 and shown in Figures 8.8 to 8.11. (Copyright ASTM. Reprinted with permission.)

provides a comparative wealth of information about the mass spectral behavior of this family of compounds. In addition, a study of cocaine by high-resolution mass spectrometry was carried out by Shapiro and co-workers in 1983, providing empirical formulas for the major ions in the spectrum.

Correlating fragment ions in the metabolite and derivative spectra with their corresponding ions in the spectrum of cocaine shows how the masses of fragment ions change in response to changes in attached functional groups. In this way various parts of the molecule can be followed through fragmentation processes in much the same manner as the deuterated derivatives of 2-pentanone and cyclohexanone in Chapter 6. When functional groups differ significantly from those in the parent compound, however, these correlations are more difficult to make. For example, the spectra of cocaethylene, *p*-hydroxycocaine, and hydroxymethoxycocaine (Figure 8.6) appear to have a peak-for-peak correspondance to that of cocaine. On the other hand, the spectra of methyl ecgonine, benzoylecgonine, and norcocaine (Figure 8.7) are quite different from that of cocaine, making it hard to decipher which peaks correspond with each other.

Mass spectral correlations for a number of cocaine metabolites, as well as the empirical formulas determined by Shapiro et al. (1983), are given in Table 8.1. This table also includes data for a deuterated derivative, *N*-trideuteriomethylnorcocaine (*d*₃-cocaine), an internal standard used for GC/MS quantitation of cocaine by selected ion

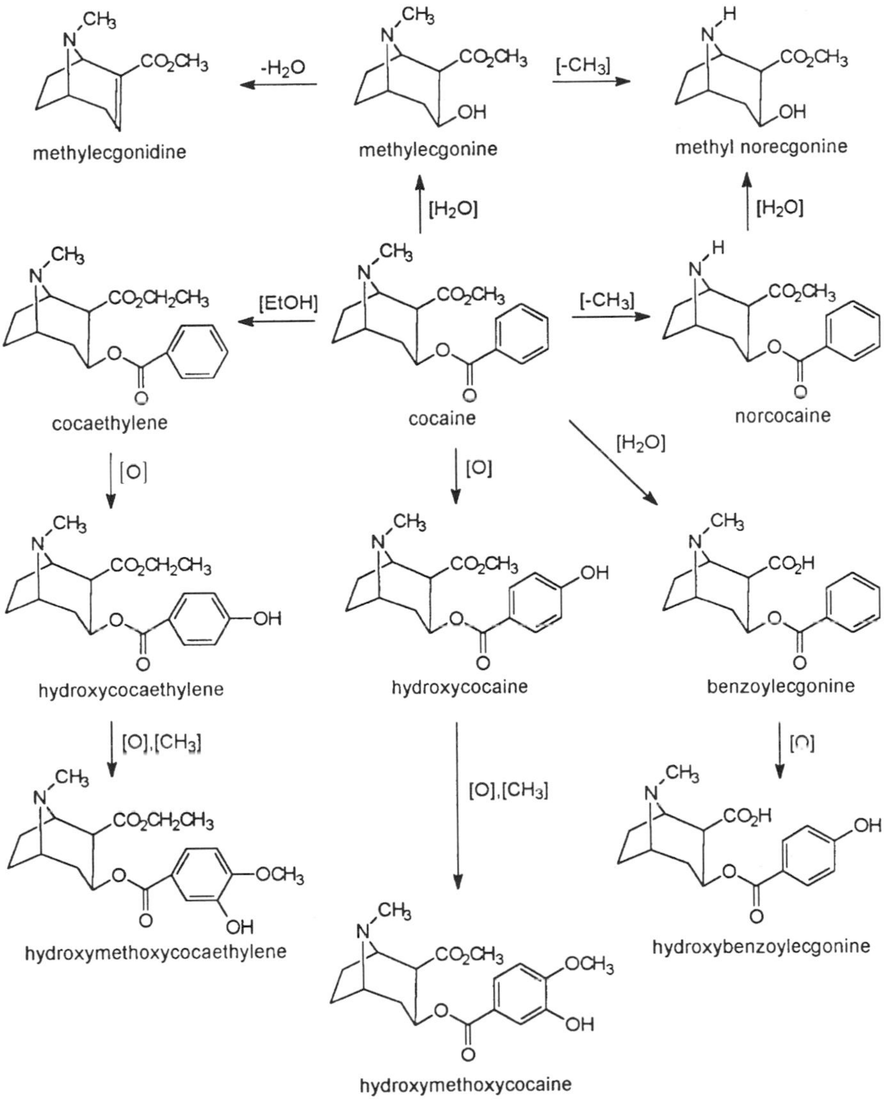

**Figure 8.5.** Cocaine and most of its known metabolites. All of these compounds have been characterized by mass spectrometry.

monitoring. Some preliminary comments need to be made about this table. First, these compounds differ from one another by substitution at three different sites: the alkyl group attached to the nitrogen ($R^1$), the alkyl ester group ($R^2$), and the aromatic ring (Ar = aryl). Second, the Roman numerals refer to structures, assigned on the basis of these correlations, to major ions in the spectrum of cocaine (Figures 8.8 to 8.11). Third, although many correlations are easy to discern, others are less clear and are denoted in the table by question marks. Dashes in the table signify the absence of ions

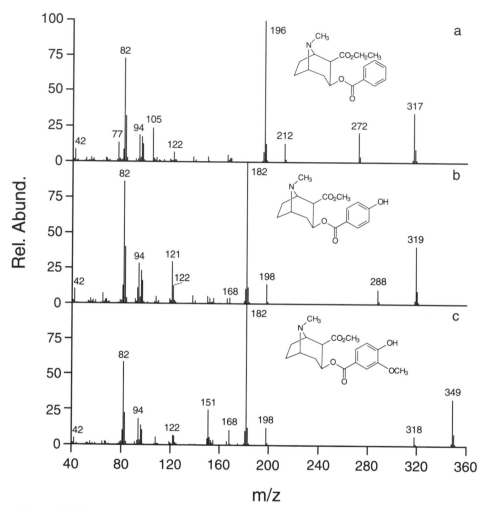

**Figure 8.6.** Mass spectra of three cocaine metabolites: (*a*) cocaethylene, (*b*) *p*-hydroxyco-caine, and (*c*) hydroxymethoxycocaine. In contrast to those in Figure 8.7, the spectra of these compounds are visually very similar to that of cocaine.

either because appropriate functional groups are not present in the derivative or because the ion is so small that it was not observed. Finally, ions for which no correlations could be determined are not represented in this table.

Let us examine the data in Table 8.1, looking specifically at several different types of derivatives. Replacing the *N*-methyl group of cocaine with a $CD_3$ group provides a unique tool since the spectrum of this compound should be virtually identical with that of cocaine except for the displacement of some peaks in the spectrum by 3 daltons (or possibly fewer, if certain rearrangement processes are involved). Indeed what we find in the spectrum of $d_3$-cocaine is that only three of the ions listed in Table 8.1 *do not move* when compared to the spectrum of cocaine. Furthermore, all of the shift-

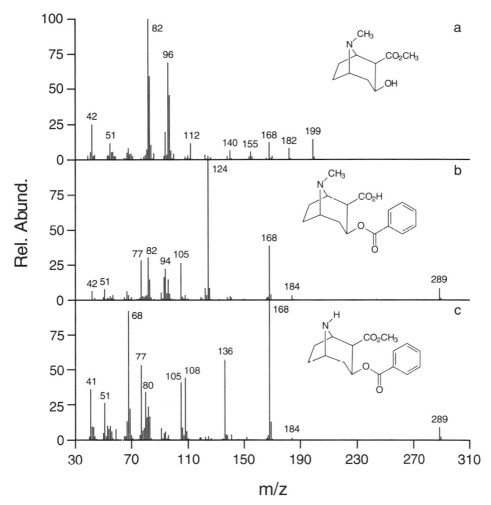

**Figure 8.7.** Mass spectra of three cocaine metabolites: (*a*) methyl ecgonine, (*b*) benzoylecgonine, and (*c*) norcocaine. Although these spectra have peaks in common with that of cocaine, correlation of peak movement to functional group changes is not straightforward.

ed peaks move by 3 daltons, indicating that the *N*-methyl group (and presumably the nitrogen itself) are retained in those fragments. This constitutes most of the major ions in the spectrum and reflects, as discussed repeatedly, the propensity for aliphatic nitrogen to direct fragmentation and to stabilize the positive charge. The only ions that do not shift are one of the ions at *m/z* 122 (see below) and those in the benzoyl low mass ion series at *m/z* 105, 77, and 51.

Close examination of the spectrum of the trideuterio derivative reveals an ion at *m/z* 125 (IX) that corresponds to *m/z* 122 in the cocaine spectrum. Although a small residual peak remains at *m/z* 122, which theoretically could correspond to the ion at *m/z* 119 in the cocaine spectrum, the spectra of other derivatives (below) also identi-

**Table 8.1. Prominent Ions in the Mass Spectra of Substituted Cocaines**

| | Functional Group | | | I | II | III | IV | V | VI |
|---|---|---|---|---|---|---|---|---|---|
| | $R^1$ | $R^2$ | Ar | $C_{17}H_{21}NO_4$ | $C_{16}H_{18}NO_3$ | — | $C_{10}H_{16}NO_3$ | $C_{10}H_{16}NO_2$ | $C_9H_{12}NO_2$ |
| Cocaine | $CH_3$ | $CH_3$ | φ | 303 | 272 | 244 | 198 | 182 | 166 |
| $D_3$-Cocaine | $CD_3$ | $CH_3$ | φ | 306 | 275 | — | 201 | 185 | 169 |
| Norcocaine | H | $CH_3$ | φ | 289 | — | — | 184 | 168 | ? |
| Benzoylecgonine | $CH_3$ | H | φ | 289 | 272 | — | 184 | 168 | ? |
| Cocaethylene | $CH_3$ | $CH_3CH_2$ | φ | 317 | 272 | 244 | 212 | 196 | 166 |
| Arylhydroxycocaine | $CH_3$ | $CH_3$ | b | 319 | 288 | 260 | 198 | 182 | 166? |
| Hydroxymethoxycocaine | $CH_3$ | $CH_3$ | c | 349 | 318 | — | 198 | 182 | 166? |
| Hydroxycocaethylene | $CH_3$ | $CH_3CH_2$ | b | 333 | 288 | 260 | 212 | 196 | 166? |

*Empirical formulas from Shapiro *et al.* (1983).
b Ar = $C_6H_4OH$.
c Ar = $C_6H_3(OMe)(OH)$.
Copyright ASTM. Reprinted with permission.

fy an ion at *m/z* 122 that contains no nitrogen—namely benzoic acid radical ion (X). The high-resolution mass spectrum of cocaine corroborates the presence of two ions at this mass (Table 8.1).

*m/z 272.* The spectrum of this single derivative offers few clues about the structures of any of these ions except that most of them contain nitrogen. More information is obtained by studying the spectrum of cocaethylene, a metabolite in which the methyl ester group has been exchanged for an ethyl group (compare the data in Table 8.1 with the spectrum for this compound in Figure 8.6a). In this spectrum ion II, which occurs at *m/z* 272 for cocaine, does not shift. This is consistent with the high-resolution data and indicates loss of the alkoxy radical of the $R^2$ (alkyl) ester group by α-cleavage (see Problem 5.7i). Indeed the spectrum of *any* cocaine derivative in which only the methoxy group is replaced by another ester group should produce an ion at *m/z* 272 [Eq. (8.5) and Figure 8.8]:

(8.5)

I (m/z 303)          II (m/z 272)          III (m/z 244)

Empirical Formula[a]

| VII $C_8H_{10}NO_2$ | VIII $C_9H_{12}NO$ | IX $C_8H_{12}N$ | X $C_7H_6O_2$ | XI $C_7H_{10}N$ | XII $C_7H_5O$ | XIII $C_6H_{10}N$ | XIV $C_6H_8N$ | XV $C_5H_8N$ | XVI $C_6H_5$ | XVII — |
|---|---|---|---|---|---|---|---|---|---|---|
| 152–155 | 150 | 122 | 122 | 108 | 105 | 96–97 | 94 | 82 | 77 | 42 |
| 155–158 | 153 | 125 | 122 | 111 | 105 | 99–100 | 97 | 85 | 77 | 45 |
| 138–141? | 136? | ? | ? | 94? | 105 | 82–83 | 80 | 68 | 77 | — |
| ? | 150? | ? | 122? | 108 | 105 | 96–97 | 94 | 82 | 77 | 42 |
| 166–169 | 150 | 122 | 122 | 108 | 105 | 96–97 | 94 | 82 | 77 | 42 |
| 152–155 | 150 | 122 | 138 | 108 | 121 | 96–97 | 94 | 82 | 93 | 42 |
| 152–155? | 150? | 122 | 168 | 108 | 151 | 96–97 | 94 | 82 | 123 | 42 |
| 166–169? | 150 | 122? | 138 | ? | 121 | 96–97 | 94 | 82 | 93? | 42 |

*m/z 244.* Another high mass ion that does not shift in the spectrum of cocaethylene is *m/z* 244 (III). Since the difference in mass between *m/z* 272 and 244 is 28 daltons, the methyl ester carbonyl seems likely to account for this loss by analogy with fragmentations of other carbonyl compounds [e.g., Eq. (4.7)]. Formation of an unconjugated secondary carbocation by simple loss of CO from II, however, is not expected. On the other hand, "nucleophilic displacement" of CO by one of the lone pairs of electrons on the benzoyl oxygen forms an ion that has the positive charge stabilized by both of the remaining oxygens (Figure 8.8).

*m/z 198 and 182.* Ions IV and V shift to higher mass by 14 daltons in the spectrum of cocaethylene, indicating that, in contrast to II and III, both of these ions retain the $R^2$ ester group. A different perspective is gained from the spectrum of *p*-hydroxycocaine (Figure 8.6*b*), in which the aromatic ring now carries the extra functional group. In this spectrum, ions retaining the aromatic ring shift to higher mass by 16 daltons (the mass of the added oxygen). As expected, ions II and III move, confirming that formation of these ions does not involve losses from the aromatic ring. More instructive, however, is the fact that ions IV and V remain at *m/z* 198 and 182, indicating that both of these ions lose the aromatic ring during their formation. Formation of *m/z* 198 by simple charge migration α-cleavage at the benzoyl group seems unlikely since it would leave the positive charge isolated on the remaining oxygen without assisting stabilization:

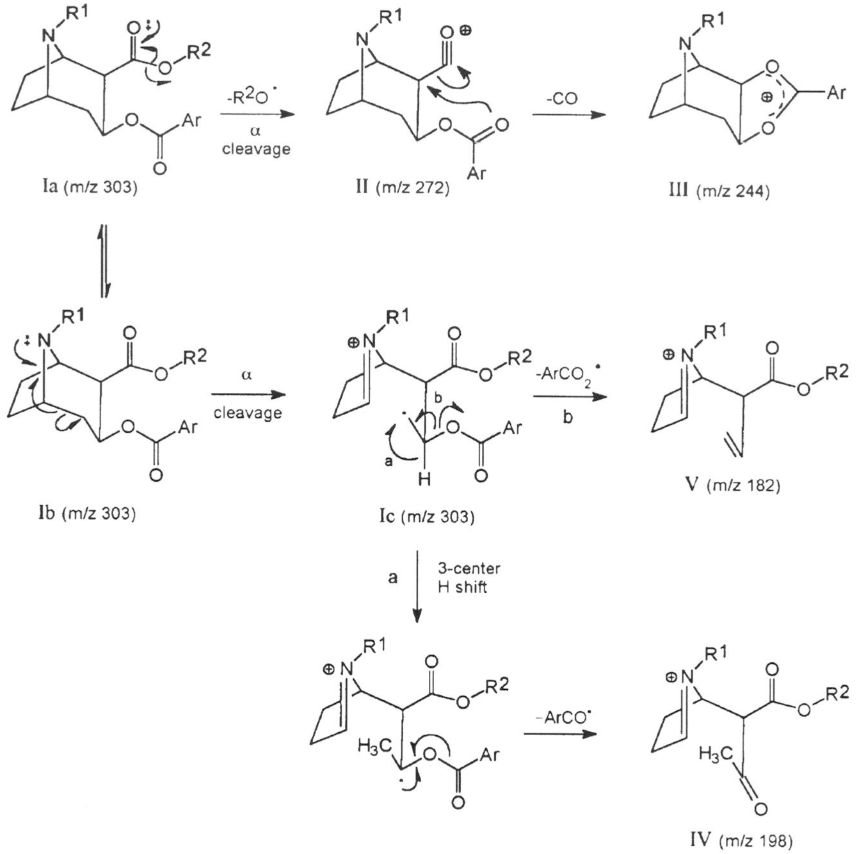

**Figure 8.8.** Primary high mass fragmentations of cocaine. (Copyright ASTM. Reprinted with permission.)

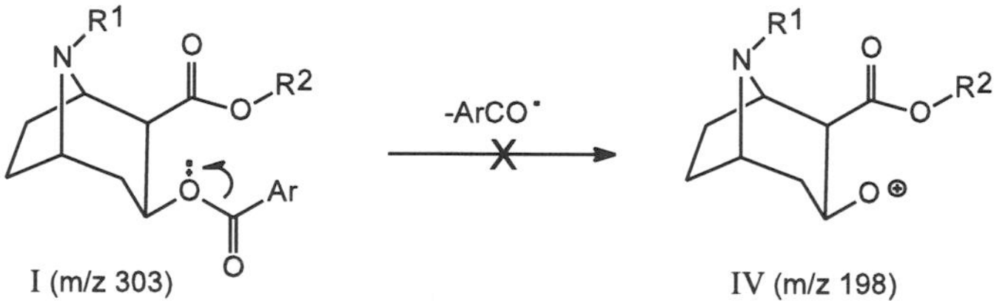

Further, $m/z$ 182, 16 daltons lower than $m/z$ 198, must result from loss of benzoate radical, but again, simple cleavage results in breaking a carbon–heteroatom bond and forms a secondary carbocation with no further conjugative interactions. That such a process should lead to one of the more intense ions in the spectrum should seem counterintuitive by now:

I (m/z 303)  (8.7)  V (m/z 182)

Indeed, we should stop now and remind ourselves that, in aliphatic amines, nitrogen nearly always directs the fragmentation. Thus, instead of considering only α-cleavage at the ester oxygens, let us focus on what might happen if initial ionization occurred at nitrogen. In that case, as with cyclohexanone (Section 6.3) and phencyclidine (Problem 6.3), α-cleavage next to nitrogen produces no immediate loss of any fragments, but rather only new molecular ions [Eq. (8.8)] that must break additional bonds before evidence of their formation can be detected by mass spectrometry.

(8.8)

One of these ions looks particularly inviting as a prospect for subsequent loss of both benzoyl and benzoate [Eq. (8.9) and Figure 8.8] since both of the resultant ions have the positive charge stabilized by nitrogen.

IV (m/z 198)  (8.9)  V (m/z 182)

Although the structure for ion V has much appeal, two alternative structures for ion IV are shown in Eq. (8.9) and Figure 8.8, both of which are consistent with all of the data and still retain the positive charge on nitrogen.

*m/z 122 (two ions) and 105.* At the low mass end of the hydroxycocaine spectrum, ions XII and XVI (the benzoyl ion series) have shifted to *m/z* 121 and 93, in keeping with the additional oxygen on the aromatic ring. Also, there are ions at both *m/z* 122 and 138; the former corresponds to ion IX, and, since it does not shift in the spectra of either cocaethylene or hydroxycocaine, must contain neither the carboalkoxy or aromatic ester groups. The empirical formula for IX bears this out. The latter ion (X) corresponds to hydroxybenzoic acid radical ion. Further confirmation of this assignment can be seen in the spectrum of hydroxymethoxycocaine, in which the benzoyl/benzoic acid ion series shifts to $m/z$ 123, 151, and 168 while ion IX remains at $m/z$ 122. Formation of the benzoic acid radical ion can be rationalized by initial ionization at the ester oxygen and four-center hydrogen migration:

$$(8.10)$$

This process, which has ample precedent, forms an ion of stability similar to that of the starting ion by ejection of an olefin.

*Other Ions.* The remaining fragmentations of the cocaine skeleton [Figures 8.9 to 8.11 and Eq. (8.11); Equation (8.11) copyright ASTM, reprinted with permission] are less straightforward.

$$(8.11)$$

The structures of ions VII, XIII, and XIV cannot be determined with certainty, and two different mechanisms for their formation are shown. Note that the first step proposed in each of these mechanisms is loss of an olefin with formation of an ion having stability similar to that of the parent ion (Figures 8.10 and 8.11). In contrast, the formation of ion XV [Eq. (8.11)] is driven by formation of a new $\sigma$-bond, extended conjugation in the ion product, and continued stabilization of the positive charge on nitrogen.

Although the fragmentations mechanisms shown in Figures 8.8 to 8.11 are consistent with the data in Table 8.1, lead to ions of reasonable stability, and are examples or extensions of familiar types of fragmentation processes, they are by no means "proven." Particularly intriguing are the formation of ion VI ($m/z$ 166), which appears to contain neither the $R_2$ group nor the aromatic ring (Table 8.1 and Figure 8.9), and ion VIII ($m/z$ 150), which is structurally unrelated to the group of ions at $m/z$ 152 to 155 (Figures 8.9 and 8.10). It is important to remember that complex molecules sometimes undergo substantial, even apparently implausible, rearrangement on their way

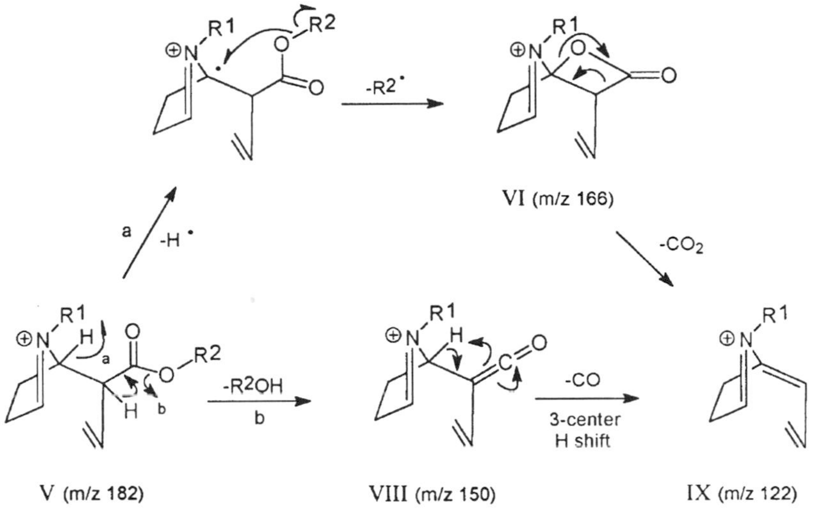

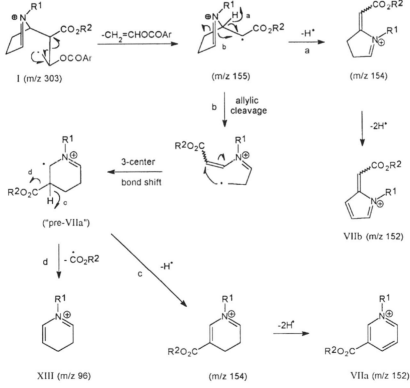

**Figure 8.9.** Proposed fragmentations of ion V ($m/z$ 182 in cocaine) to form VI, VIII, and IX. (Copyright ASTM. Reprinted with permission.)

**Figure 8.10.** Proposed secondary fragmentations of the molecular ion of substituted cocaines after initial α-cleavage. (Copyright ASTM. Reprinted with permission.)

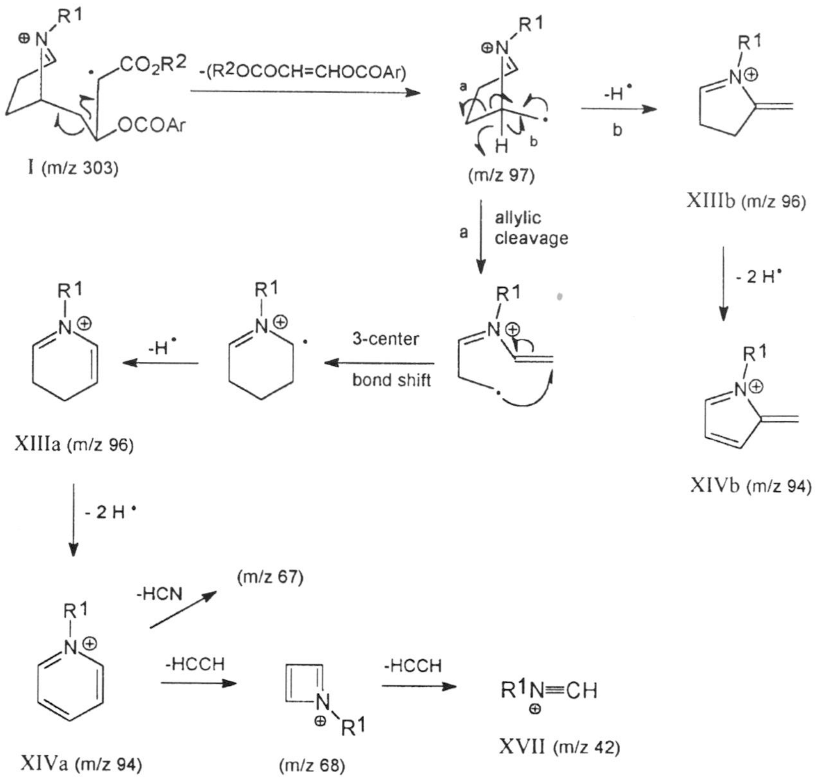

**Figure 8.11.** More proposed secondary fragmentations of the molecular ion of substituted cocaines after initial α-cleavage. (Copyright ASTM. Reprinted with permission.)

to forming stable low mass ions. Because they tend to lose their original structural integrity, these ions are usually of less value in structure determination.

*Application.* The fragmentation schemes developed from the data in Table 8.1 can be used to predict the mass spectrum of phenylacetylmethylecgonine:

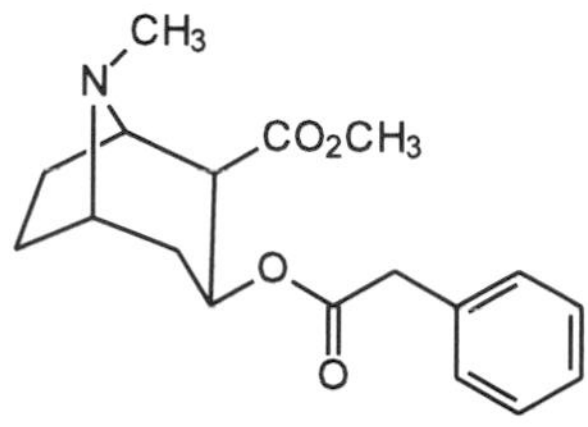

This compound is isomeric with cocaethylene but has the extra methylene group between the phenyl ring and the "benzoyl" carbonyl group. We will assign approximate peak intensities to the various fragments (and their isotopic abundance ions) based on the intensities of corresponding ions in the spectrum of cocaine.

The molecular ion for this compound will occur at $m/z$ 317. Since the $R^2$ group in this molecule is methyl, ion II will occur at $m/z$ 286 [$(M - 31)^+$] from loss of a methoxy group and ion III at $m/z$ 258 [$(M - 31-28)^+$] from the subsequent loss of CO. Because ions IV and V involve loss of the aromatic ring and its attached carbonyl group, they should remain at $m/z$ 198 and 182, respectively. Ions VI to VIII all have lost the aromatic ring (Figures 8.9 and 8.10), so that these ions should not shift from their positions in the cocaine spectrum.

Although ion IX, having lost both ester groups, stays at $m/z$ 122, ions X, XII, and XVI, which constitute the aromatic acid and its fragments, should all shift. The acid radical ion (X) will move to $m/z$ 136, while the phenylacetyl ion (XII) will appear at $m/z$ 119. What happens beyond this point, however, is less clear since $m/z$ 119 should lose CO to form the benzyl ion ($m/z$ 91), rather than phenyl ($m/z$ 77), and the benzyl ion has its own ion series at $m/z$ 65 and 39. Exactly how energy will be apportioned in the fragmentation of these ions is unpredictable, so that the actual relative intensities of these ions may differ from those expected.

Finally, the major nitrogen-containing ion clusters at $m/z$ 94 to 97 (XIII and XIV), $m/z$ 82 and 83 (XV), and $m/z$ 42 (XVII) will remain at these masses since only the unchanged core of the original molecule is left in these ions.

Figure 8.12 shows the results of these predictions and an actual spectrum of phenylacetylmethylecgonine for comparison. Although there are several minor differences between the spectra (and there should be since not all of the ions from the cocaine spectrum were even considered), the predicted spectrum is remarkably accurate.

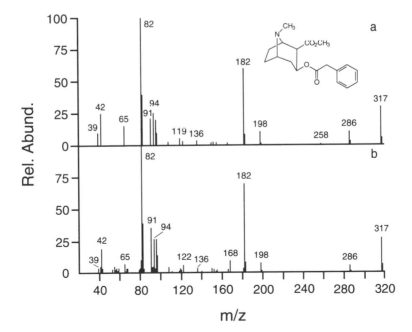

**Figure 8.12.** (*a*) Mass spectrum of phenylacetylmethylecgonine predicted by reference to fragmentations of cocaine derivatives (see text), and (*b*) actual mass spectrum of phenylacetylmethylecgonine.

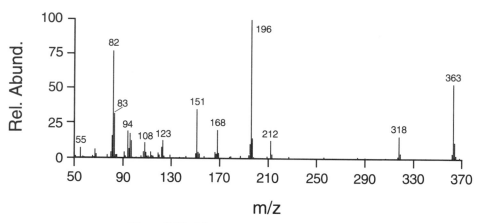

**Figure 8.13.** Mass spectrum for Problem 8.3.

**Problem 8.2.** What differences would you expect between the spectra of phenylacetylmethylecgonine (Figure 8.12) and that of toluylmethylecgonine (structure below), a recently synthesized cocaine analog?

**Problem 8.3.** Using the information in this section, identify the compound that gave rise to the spectrum in Figure 8.13.

## 8.4. PHENCYCLIDINE AND ITS ANALOGS

### 8.4.1. Fragmentations of Phencyclidine

Phencyclidine is a veterinary tranquilizer with a powerful and dangerous hallucinogenic effect in humans. Also known as angel dust and PCP (an acronym for the chemical name 1-phenylcyclohexylpiperidine), this drug and its analogs have enjoyed sporadic periods of illicit popularity for over two decades. At first glance, the mass spectrum of phencyclidine (Figure 8.14) is surprising. Facile hydrogen loss usually occurs only from highly activated positions—for example, on carbons directly attached to two aromatic rings (see Section 5.2). The prominent loss of a propyl group (*m/z* 200) seems unusual, given the lack of aliphatic groups in the molecule.

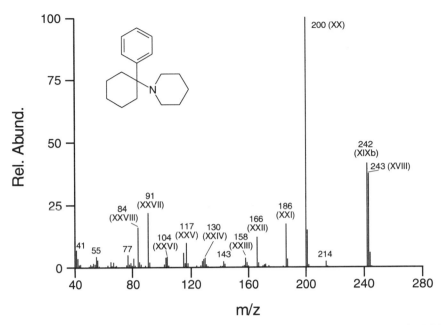

**Figure 8.14.** Mass spectrum of phencyclidine. Roman numerals refer to ion structures in Table 8.2 and Figures 8.15 to 8.17.

Although the mass spectra of several analogs are available, chemists at the Drug Enforcement Administration studied the mass spectra of phencyclidine and three of its deuterated derivatives to gain insight into the fragmentation of this molecule (Clark, 1986). The results of this study are summarized in Table 8.2 (as in Table 8.1, the Roman numerals refer to ions in Figure 8.14 and their corresponding structures in Figures 8.15 to 8.17). The derivatives used were those in which (a) all of the aromatic hydrogens had been replaced with deuterium (denoted here as the $d_5$ derivative), (b) all of the cyclohexyl hydrogens were replaced with deuterium (the $d_{10}$ derivative), and (c) all of the hydrogens on both the phenyl and cyclohexyl rings were replaced (the $d_{15}$ derivative). The $d_5$ derivative was synthesized by initially combining $d_5$-bromobenzene (after reaction with magnesium to form the Grignard reagent) with cyclohexanone. The $d_{10}$ derivative was made by using decadeuteriocyclohexanone as one of the starting materials.

*m/z 242.* Table 8.2 contains an unexpected result. First, in the spectrum of the $d_5$ derivative, ion XIX, corresponding to the loss of hydrogen radical from the molecular ion, shifts not from $m/z$ 242 to 247 (as expected if the hydrogen is lost via $\alpha$-cleavage from one of the carbons next to the nitrogen in the piperidine ring), but rather to $m/z$ 246. The intensity of the peak at $m/z$ 247 in this spectrum is consistent only with the $^{13}C$ contribution from the $m/z$ 246 ion. There is only one possible interpretation to this data: The lost hydrogen comes from the phenyl ring! This loss seems so unusual that some convincing mechanism must be devised to account for its occurrence.

**Table 8.2. Prominent Ions in the Mass Spectra of Some Substituted Phencyclidines**

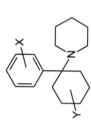

| Name | X | Y | XVIII | XIX | XX | XXI | XXII | XXIII | XXIV | XXV | XXVI | XXVII | XXVIII |
|---|---|---|---|---|---|---|---|---|---|---|---|---|---|
| Phencyclidine | — | — | 243 | 242 | 200 | 186 | 166 | 158 | 130 | 117 | 104 | 91 | 84 |
| $d_5$-Phencyclidine | $d_5$ | — | 248 | 246 | 205 | 190,191 | 166 | 163 | 135 | 122 | 109 | 96 | 84 |
| $d_{10}$-Phencyclidine | — | $d_{10}$ | 253 | 252 | 203 | 188 | 176 | 167 | 135 | 120+ | 104,105 | 92,93 | 85 |
| $d_{15}$-Phencyclidine | $d_5$ | $d_{10}$ | 258 | 256 | 208 | 192,193 | 176 | 172 | 140 | 125+ | 109,110 | 97,98 | 85 |
| Arylmethyl-PCP | $CH_3$ | — | 257 | 256[a] | 214 | 200 | 166 | 172 | 144 | 131 | ? | 105 | 84 |

[a]*ortho*-Methylphencyclidine also shows a significant (10%) ion at mass 242 from loss of the aryl methyl group.

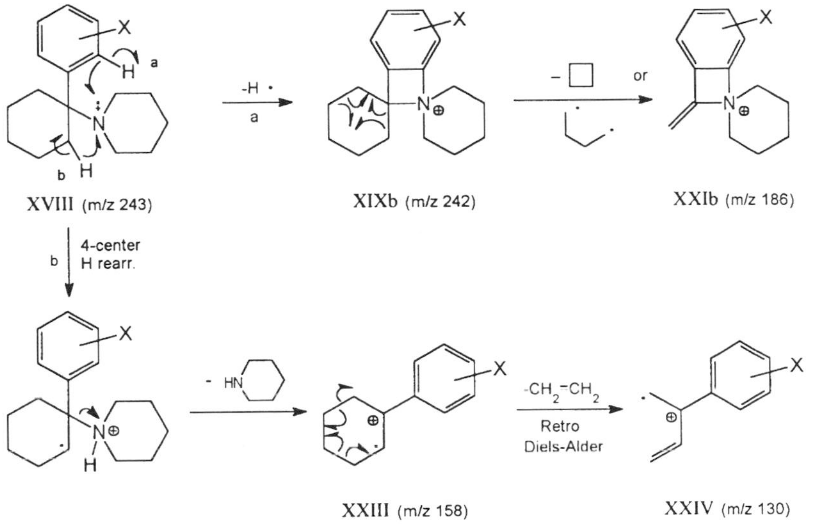

**Figure 8.15.** Fragmentation of phencyclidine showing losses of a phenyl(!) proton and piperidine.

It can indeed be rationalized if, after initial ionization at nitrogen, one of the *ortho* hydrogens on the phenyl ring is lost and a carbon–nitrogen bond is formed that retains the positive charge on nitrogen [Eq. (8.12) and Figure 8.15):

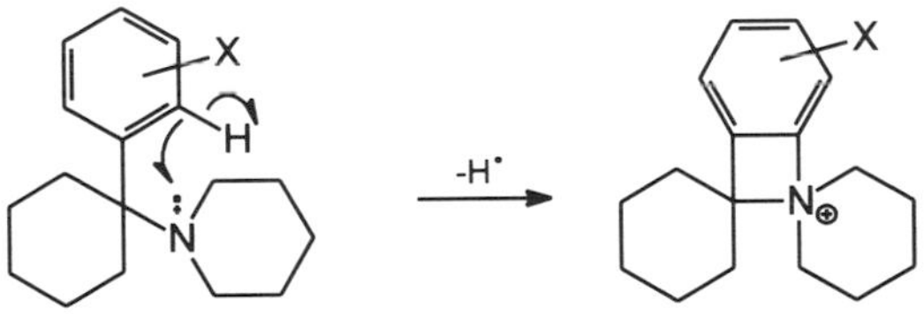

(8.12)

XVIII (m/z 243)        XIXb (m/z 242)

This behavior is similar to that exhibited by *o*-ethoxybenzamide in the previous chapter. Support for this mechanism comes from the mass spectra of arylmethyl analogs of phencyclidine, in which the *meta*- and *para*-isomers show virtually no loss of the arylmethyl group (as expected), but the *ortho* isomer produces a 10% peak corresponding to the loss of this group [Lodge et al., 1992; Eq. (8.13)]:

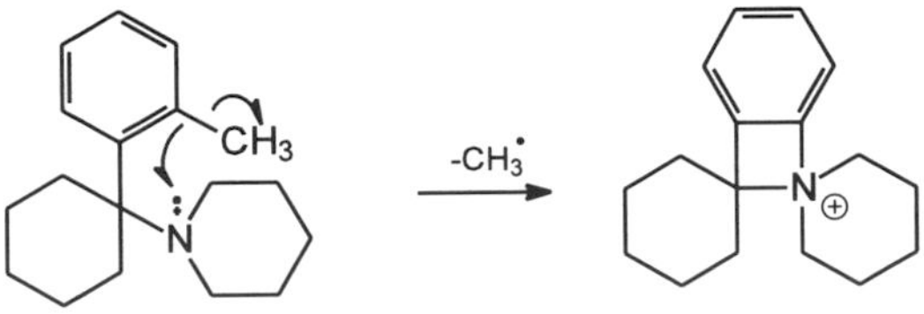

(8.13)

m/z 242

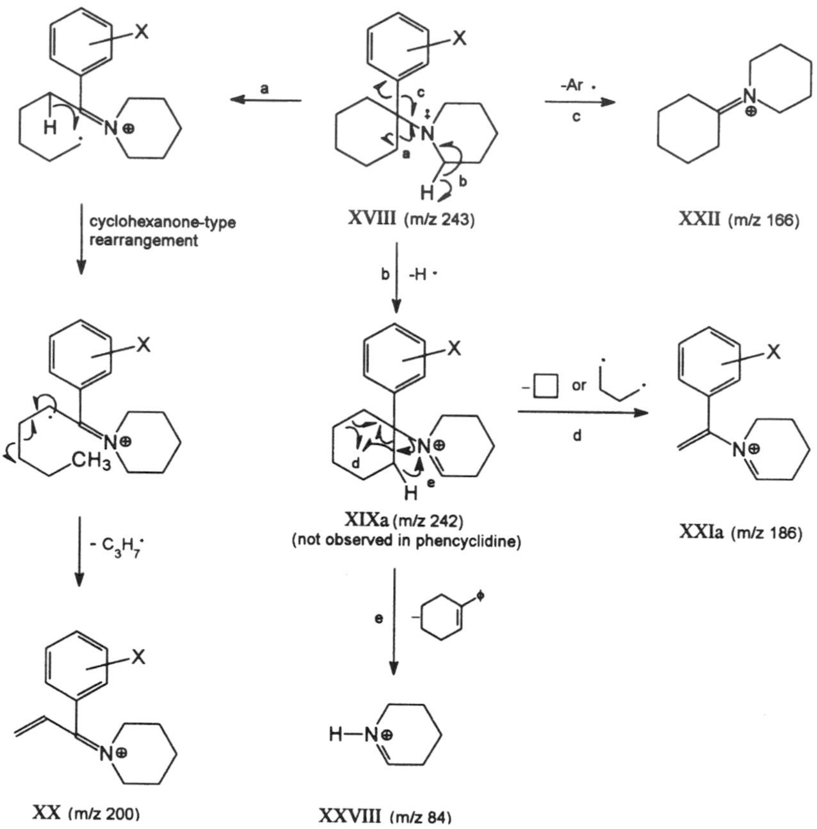

**Figure 8.16.** Some primary fragmentation modes of phencyclidine. Note: Ar = aryl.

The remaining data for the m/z 242 ion are also consistent with this interpretation since the $d_{10}$ derivative shows no loss of deuterium and the $d_{15}$ derivative loses only one.

*m/z 200.* Formation of the ion at m/z 200 (XX) via a cyclohexanone-type rearrangement (see Problem 6.3) is supported by the deuterium labeling data. In particular, the deuterated aromatic ring of the $d_5$ derivative shows no loss of deuterium, while the $d_{10}$ derivative loses 7 of the 10 deuteriums in the cyclohexane ring; the same pattern is reflected in the spectrum of the $d_{15}$ compound. Such losses are expected if the propyl group ($C_3H_7$) is lost entirely from the cyclohexane ring (Figure 8.16). The small ion at m/z 214 results from loss of an ethyl group, rather than a propyl group, from the radical ion intermediate in this rearrangement [Eq. (8.14); compare Figure 6.11):

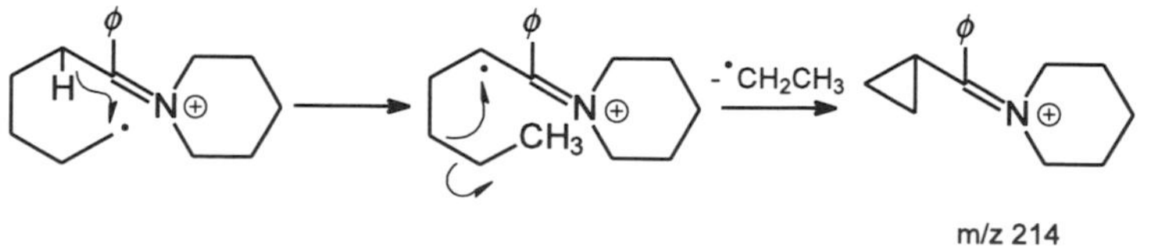

$$(8.14)$$

m/z 214

*m/z 186.* Ion XXI (*m/z* 186 in the phencyclidine spectrum) appears as a pair of ions in the spectra of the $d_5$ and $d_{15}$ derivatives! In the spectrum of the $d_5$ derivative, these peaks are seen at *m/z* 190 and 191 in a ratio of approximately 4:3, indicating a nearly equal tendency to lose either one or no deuterium from the phenyl ring. At the same time, the $d_{10}$ compound loses eight deuteriums from the cyclohexane ring since this ion shifts to higher mass by only 2 daltons. For the $d_{15}$ derivative, the pair of ions appears at *m/z* 192 and 193 in an approximate ratio of 2:1, showing losses of eight cyclohexyl deuteriums *plus* either one or no phenyl deuteriums.

An explanation for this phenomenon comes from the formation of ion XIX by two different pathways—one involving the unexpected loss of a phenyl proton, the other by the predicted loss of a proton by α-cleavage from the piperidine ring:

XVIII (m/z 243)            XIXb (m/z 242)            XXIb (m/z 186)      (8.15)

XVIII (m/z 243)            XIXa (m/z 242)            XXIa (m/z 186)      (8.16)

This would account for the two peaks observed in the spectra of the $d_5$ and $d_{15}$ derivatives. The remaining step(s) in this fragmentation involve the loss of four carbons and eight hydrogens from the cyclohexane ring. Although the ions formed in this fragmentation (XXIa and XXIb) are undoubtedly stabilized by extended conjugation, the nature of the lost fragment is not clear. Formation of an additional σ-bond to form cyclobutane is a tempting rationalization, but a $C_4H_8$ diradical cannot be ruled out.

*m/z 166.* Ion XXII occurs at *m/z* 166 in the phencyclidine spectrum, 77 daltons below the molecular ion. Loss of phenyl radical by α-cleavage is expected in this molecule (Figure 8.16). The data in Table 8.2 bear this out since the $d_5$ derivative shows the loss of all five phenyl deuteriums, while the $d_{10}$ derivative loses no deuterium at all.

*m/z 158 and 130.* The small ion at *m/z* 158 (XXIII) retains all of the hydrogens in the phenyl ring and all but a single hydrogen from the cyclohexane ring, consistent with loss of piperidine from the molecular ion (Figure 8.15). Although such fragmentations usually do not compete well with α-cleavage, the ion formed here is stabilized by conjugation with the aromatic ring. Ion XXIII, the molecular ion of 1-phenylcyclohexene, undergoes the retro Diels–Alder fragmentation (Section 6.4) to give ion XXIV at *m/z* 130 by loss of ethylene (Figure 8.15). Consistent with this hypothesis, Table 8.2 shows that, in the spectra of the deuterated derivatives, ion XXIII loses an additional four deuteriums from the cyclohexane ring without disturbing the phenyl ring.

*m/z 84.* Ion XXVIII is the only other ion in the spectrum with a simple pattern of deuterium loss. Formation of this ion involves loss of all of the phenyl hydrogens as well as all but one of the cyclohexane hydrogens, consistent with the loss of phenylcyclohexene from the unobserved ion XIXa at *m/z* 242 via secondary rearrangement of the initially formed α-cleavage ion (Figure 8.16). Interestingly, this ion can be formed *only* via ion XIXa. Ion XIXb, which is not an α-cleavage ion, must break bonds to both the cyclohexane and aromatic rings, in addition to rearranging hydrogen, to generate ion XXVIII. The behavior of ion XXVIII in the spectra of the deuterated derivatives provides impressive evidence for the formation of ion XIXa in the fragmentation of phencyclidine *despite the fact that no peak corresponding to the ion itself is observed.* Since ion detection is determined in part by the relative activation energies of further fragmentation reactions (and thus to the lifetime of that ion in the ion source; Chapter 3), these results indicate that ion XIXa reacts too rapidly to reach the electron multiplier, but that higher activation energies for further fragmentation of ion XIXb allow it to be detected.

*Other Ions.* The ions at *m/z* 117, 104, and 91 give patterns of deuterium loss that are actually more complex than the data in Table 8.2 indicate, implying that all three are formed via several different pathways and that their stabilities probably play a more important role in their formation than the mechanisms by which they are formed. Ion XXV, which arises primarily via loss of seven hydrogens from the cyclohexane ring while retaining the aromatic hydrogens, could have either of the two structures shown in Figure 8.17, provided both carbons in the aziridine ring of structure XXVa and the hydrogen on the nitrogen originate on the cyclohexane ring. The ion at *m/z* 115 is probably related structurally to *m/z* 117 by loss of additional hydrogen [see, e.g., Eq. (7.5)].

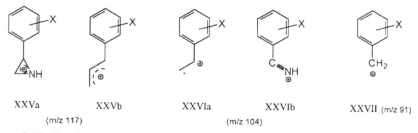

**Figure 8.17.** Possible structures for three low mass ions in the mass spectrum of phencyclidine.

Although the ion at $m/z$ 104 can also have at least two structures, XXVIa (which might seem more likely on a purely intuitive basis) is not consistent with the deuterium-labeling data. Barring migration of hydrogen *from* the piperidine ring to the cyclohexane ring (something not observed for any of the other ions), this structure should result from loss of 7 of the cyclohexane hydrogens. Instead, the most prominent in the cluster of peaks ascribable to this ion in the spectrum of the $d_{10}$ derivative show the loss of 9 or 10 cyclohexane hydrogens, more consistent with structure XXVIb (Figure 8.17). Note that this ion has the positive charge stabilized by nitrogen.

Finally, ion XXVII at $m/z$ 91 (the benzyl ion) arises primarily via the loss of eight cyclohexyl hydrogens, with a small contribution (10 to 15%) from the loss of a phenyl hydrogen *and* eight cyclohexane hydrogens. This behavior is similar to that of ion XXI, which was formed via pathways involving each of the $m/z$ 242 ions. Although the two hydrogens that end up on the benzylic carbon come from the cyclohexane ring, mechanisms accounting for these rearrangements are not straightforward.

## 8.4.2. Phencyclidine Analogs

In contrast to the spectra of cocaine and many of its derivatives, spectra of the analogs of phencyclidine do not appear to be that similar to one another (Figure 8.18). This lack of similarity, however, arises only because none of the major ions in the spectra occur at the same masses. On the other hand, when the pattern of losses from the molecular ions is considered for these compounds (Table 8.3), a different picture emerges. For example, two of the three compounds shown exhibit a prominent $(M - 1)^+$ ion [the thiophene analog, Figure 8.18c, has only one *ortho* proton to lose from the aromatic ring and thus has a relatively smaller $(M - 1)^+$ ion]. In addition, all three compounds show a small but detectable $(M - 29)^+$ ion from loss of ethyl, a base peak at $(M - 43)^+$ due to loss of propyl, an ion of moderate abundance at $(M - 57)^+$ [actually $(M - 1-56)^+$; Figures 8.15 and 8.16] and $\alpha$-cleavage loss of the aryl group. In addition the phencyclidine "low mass ion series" at $m/z$ 91, 104, 115, and 117 is reproduced nicely in Figures 8.18a and b. In the spectrum of the thiophene analog, this series is displaced to higher mass by 6 daltons, reflecting the higher mass of the thiophene ring.

The thiophene analog loses the piperidine ring much more readily than either phencyclidine or the other analogs. Since initial ionization in this compound occurs as readily at sulfur as nitrogen, loss of the piperidine ring occurs by stabilization of the positive charge on sulfur (which is less electronegative than nitrogen) via long-distance $\alpha$-cleavage:

(8.17)

m/z 249                              m/z 165

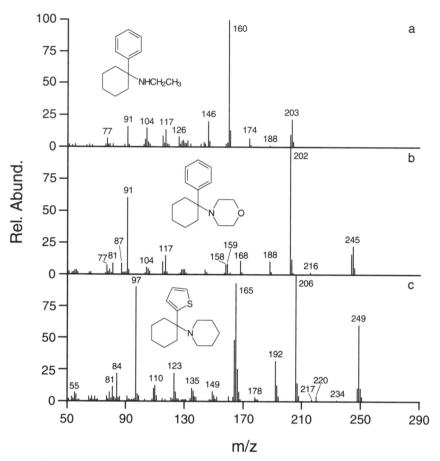

**Figure 8.18.** Mass spectra of three phencyclidine analogs: (*a*) *N*-ethyl-1-phenylcyclohexyl-amine, (*b*) *N*-(1-phenylcyclohexyl)morpholine, and (c) *N*-[1-(2-thienyl)cyclohexyl]piperidine.

**Table 8.3. Pattern of Fragmentation Losses in Phencyclidine Analogs**

| Ion | m/z in a[a] | m/z in b[a] | m/z in c[a] |
|---|---|---|---|
| M$^{+\cdot}$ | 203 | 245 | 249 |
| (M-1)$^+$ | 202 | 244 | 248 |
| (M-43)$^+$ | 160 | 202 | 206 |
| (M-1-56)$^+$ | 146 | 188 | 192 |
| (M-Ar)$^{+b}$ | 126 | 168 | 166 |
| ArCH=CHCH$_2^{+b}$ | 117 | 117 | 123 |
| ArCH$^+$—CH$_2^{\cdot b}$ | 104 | 104 | 110 |
| ArCH$_2^{+b}$ | 91 | 91 | 97 |

[a]From mass spectra for compounds shown in Figure 8.18.
[b]Ar = Aryl.

Formation of corresponding ions at *m/z* 159 in the spectra of phencyclidine and the other two analogs must, in contrast, proceed by charge migration mechanisms.

**Problem 8.4.** Using the fragmentation patterns for phencyclidine and its analogs as a model, predict the major features of the mass spectrum of *N,N*-diethyl-1-phenylcy-clohexylamine (structure below). Predict approximate relative intensities for these ions using those in Figure 8.14 as a guide.

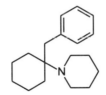

**Problem 8.5.** Identify the compound that gave rise to the spectrum in Figure 8.19.

**Problem 8.6.** In contrast to the spectra of the tolyl analogs of phencyclidine (see data in Table 8.2), the spectrum of 1-benzylcyclohexylpiperidine (structure below) has a base peak at *m/z* 166 and an ion of only 1 to 2% relative abundance at *m/z* 214 due to the cyclohexanone-type rearrangement. Explain this difference in behavior.

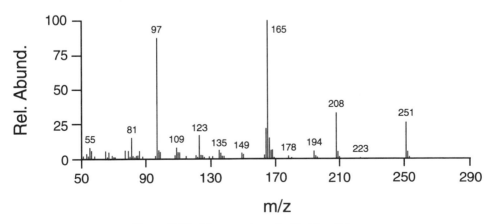

**Figure 8.19.** Mass spectrum for Problem 8.5.

## 8.5. MORPHINE ALKALOIDS

Morphine and codeine, whose mass spectra are shown in Figure 8.20, are the two principal alkaloids produced by the opium poppy. Both are strong narcotics, and many compounds that are structurally related to them have narcotic properties as well. These spectra, like those of many morphine derivatives, are dominated by molecular ions that, because of the complex interconnectivity of the ring system, cannot lose more than one or two small functional groups without substantial rearrangement.

Morphine and codeine differ by only the methyl group on the aromatic oxygen, so it is not surprising that their spectra show many similarities. In fact, these two spectra (as well as the spectra of many other morphine derivatives) are so similar below $m/z$ 170, that the ions at $m/z$ 42, 59, 70, 81, 115, 124, and 162 constitute a "morphine low mass ion series." Above $m/z$ 170, on the other hand, all of the major ions in the morphine spectrum shift to higher mass by 14 daltons in the spectrum of codeine. Thus, the important high mass ions in these spectra contain the aromatic ring and its attached oxygen, whereas the important low mass ions have lost this group (Wheeler et al., 1967). Another way of saying this is that those fragmentations involving the

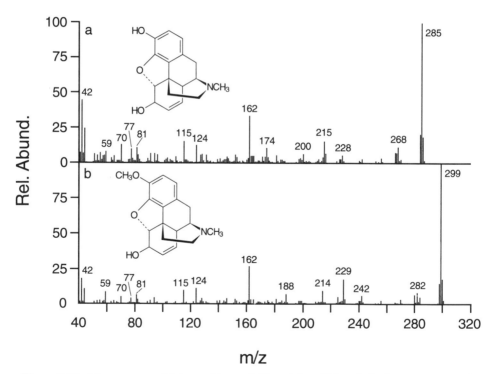

**Figure 8.20.** Mass spectra of (*a*) morphine and (*b*) codeine. Although the low mass ends of these spectra are extremely similar, nearly all of the high mass ions in the morphine spectrum are shifted to higher mass by 14 daltons in the spectrum of codeine.

fewest number of steps occur in the alicyclic portion of these molecules. In either case, we expect fragmentation to be centered on the aliphatic nitrogen—both in causing first losses from the alicyclic portion of the molecule, then in stabilizing the positive charge for ions in the low mass end of the spectrum (Figure 8.21).

The addition of fragmentable functional groups to this nucleus shifts the focus of fragmentation from the morphine ring system to the functional groups themselves. A good example is the spectrum of diacetylmorphine (heroin; Figure 8.22). The morphine low mass ion series is still visible, although the intensity of the acetyl ion ($m/z$ 43) overshadows it. The ion at $m/z$ 204 has the same structure as that of $m/z$ 162 except for the addition of the acetyl group, while the ions at $m/z$ 215 and 284 are common to both heroin and morphine.

The facile fragmentations of heroin, however, occur at high mass. The molecular ion at $m/z$ 369 loses 42 and 59 daltons to give the ions at $m/z$ 327 and 310, respectively, while $m/z$ 268 is 59 daltons less than $m/z$ 327 and 42 less than $m/z$ 310. Two types of fragmentation processes are operating here, each involving the acetyl groups. In one case, acetyl is lost as ketene (42 daltons), while in the other it is lost as acetate radical ($CH_3CO_2 \cdot$).

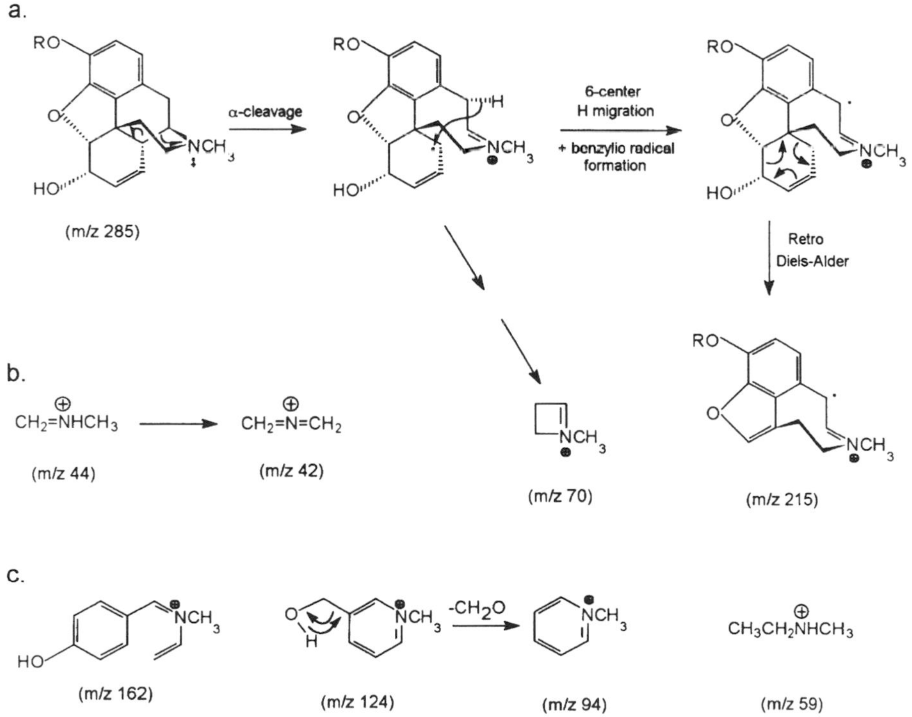

**Figure 8.21.** (*a*) Representative fragmentations of morphine; (*b*) and (*c*) structures of some low mass ions.

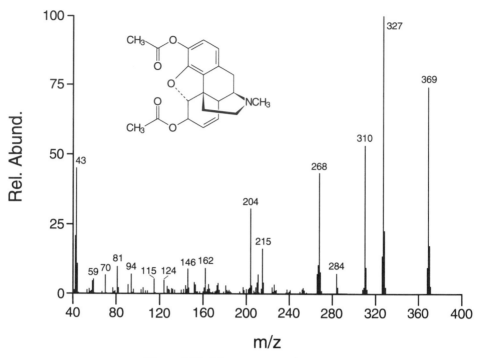

**Figure 8.22.** Mass spectrum of heroin.

Based on previous discussions, it is possible to distinguish which loss involves which acetate group. Using the behavior of acetylacetaminophen as a model [Eq. (5.20)], the acetate group attached to the aromatic oxygen should fragment via elimination of an olefin so as to maintain the aromaticity of the ring:

The nonaromatic acetate group, on the other hand, is not only allylic to the remaining double bond, but also in such a position that initial ionization *at nitrogen* can lead to loss of acetate radical with stabilization of the positive charge on the nitrogen [α-cleavage by long distance; Eq. (8.19)]:

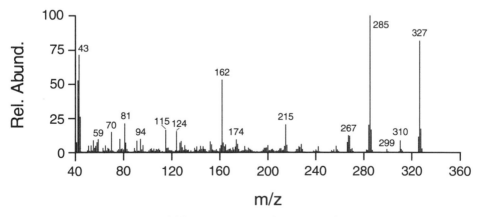

$$(8.19)$$

m/z 369                                    m/z 310

This is analogous to the loss of benzoate radical by cocaine [Eq. (8.9)]. Similar losses in the spectra of morphine and codeine produce the ions at $m/z$ 268 and 282, respectively; in these cases, however, the hydroxy radical is not as good a leaving group as acetate radical, so the ions are much less intense. The $m/z$ 268 ion in the heroin spectrum can form from either the $m/z$ 327 or 310 ion by fragmentation of the unaffected acetate group via the mechanisms just discussed.

**Problem 8.7.** Identify the compound that produced the spectrum shown in Figure 8.23.

## 8.6. TWO REAL-LIFE PROBLEMS—OPTIONAL

Most of the problems in previous sections have been limited in scope. The problems below, although they relate directly to the material covered in this chapter, have a much broader focus. They are derived from actual case samples and attempt to recreate, as well as possible, the level of information available to the analyst. Despite the fact that they relate to the analysis of illicit drugs, the problems may be solved with a knowledge only of mass spectrometry and general organic chemistry.

**Problem 8.8.** The chromatogram in Figure 8.24 resulted from the methanolic wash of a spoon that had been used for narcotics. The user apparently placed heroin (as the HCl

**Figure 8.23.** Mass spectrum for Problem 8.7.

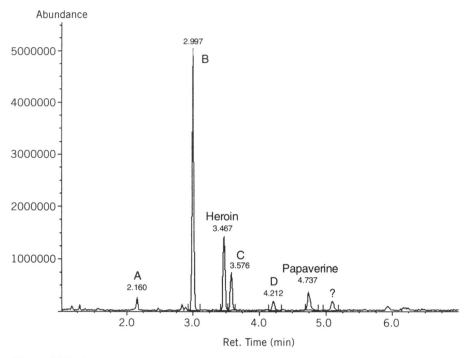

**Figure 8.24.** Chromatogram from methanolic extract of residue from a spoon used for narcotics (Problem 8.8).

salt) and a small amount of water in the spoon and heated it, intending to dissolve the heroin for injection. In fact, heroin was identified in this mixture [retention time (r.t.) 3.467 min]. Papaverine, one of the nonmorphine alkaloids of opium, also was identified in this extract (r.t. 4.747 min), indicating either that the heroin was manufactured by direct acetylation of opium or, more importantly, that opium itself was present along with the heroin. The remaining peaks in this chromatogram were unfamiliar.

The mass spectra of four of these unknowns (labeled A, B, C, and D in Figure 8.24) are shown in Figure 8.25. In determining possible structures for these compounds, the following information is helpful: (a) The major active alkaloids found in opium are morphine, codeine, and thebaine:

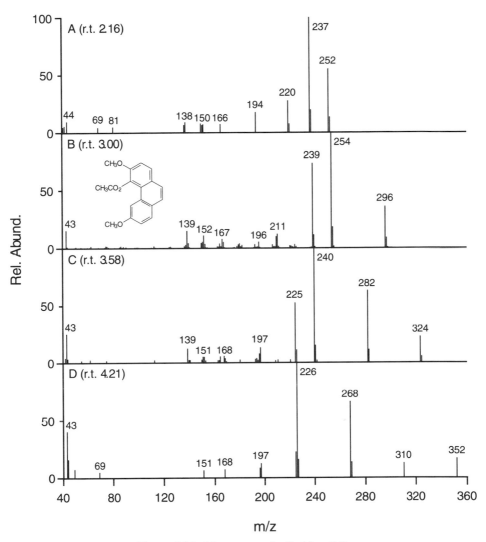

**Figure 8.25.** Mass spectra for Problem 8.8.

(b) The largest peak in the chromatogram (peak B; r.t. 2.997 min) was identified by library search as the acetyl ester of thebaol. Thebaol apparently is formed when thebaine is heated under unspecified conditions [(Eq.) 8.20]. Morphine and codeine presumably could decompose in a similar manner under the same conditions. Thermal decomposition of heroin itself might account for formation of acetylated thebaol by transacetylation.

Examine the spectrum of compound B and rationalize the major losses in terms of the proposed structure. Then study the other three spectra and devise structures that are consistent with the information given above. Remember that these spectra are rather weak and, as such, contain only a limited amount of information.

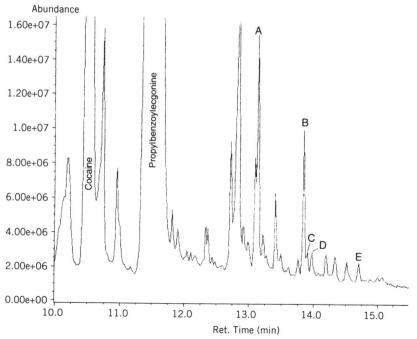

Thebaine          Thebaol          Acetylthebaol

**Problem 8.9.** The chromatogram shown in Figure 8.26 was obtained after derivatization of a urine extract designed to isolate cocaine and its metabolites. The derivatizing reagent in this case replaced all $-OH$ groups with $CH_3CH_2CH_2O-$ groups. In addition to cocaine and propylbenzoylecgonine [resulting from propylation of the cocaine metabolite benzoylecgonine; Eq. (8.21)], spectra for five other apparent cocaine metabolite derivatives were observed (labeled as peaks A, B, C, D, and E in Figure 8.26). Assign structures to these derivatives whose spectra are shown in Figure 8.27. As in the previous problem, some of these spectra are extremely weak, so that only major ions in the spectrum can be counted on to provide reliable information.

**Figure 8.26.** Partial total ion chromatogram from a urine sample extracted for basic drugs and derivatized by propylation (Problem 8.9).

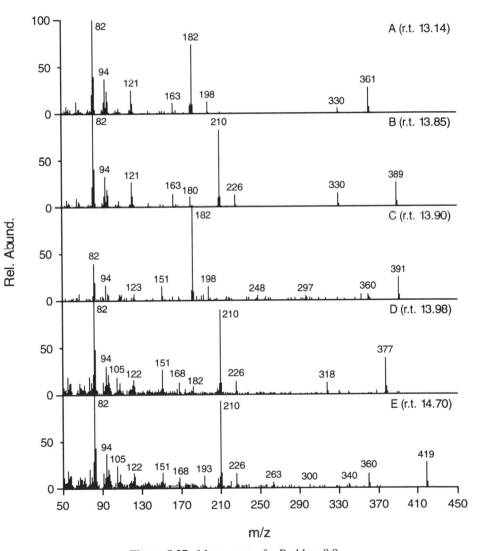

Figure 8.27. Mass spectra for Problem 8.9.

## REFERENCES

C.C. Clark, *J. Assoc. Off. Anal. Chem.,* **69,** 814–820 (1986).

B.A. Lodge, R. Duhaime, J. Zamecnik, P. MacMurray, and R. Brousseau, *Forensic Sci. Int.,* **55,** 13–26 (1992).

R.H. Shapiro, D.S. Amenta, M.T. Kinter, and K.B. Komer, *Spectroscopy: Int. J.,* **2,** 227–231 (1983).

R.M. Smith, *J. Anal. Toxicol.,* **8,** 35–37 (1984).

R.M. Smith, *J. Anal. Toxicol.,* **8,** 38–42 (1984).

R.M. Smith, *J. Forensic Sci.,* **42,** 475–480 (1997).

R.M. Smith, M.A. Poquette, and P.J. Smith, *J. Anal. Toxicol.,* **8,** 29–34 (1984).

R.M. Smith and L.A. Nelsen, *J. Forensic Sci.,* **26,** 280 (1991).

D.M.S. Wheeler, T.H. Kinstle, and K.L. Rinehart, *J. Am. Chem. Soc.,* **89,** 4494–4501 (1967).

# CHAPTER 9

# ANSWERS TO PROBLEMS

## CHAPTER 1

**1.1.** The structures proposed for these ions will be more understandable after the discussion of α-cleavage in Chapter 5.

| *m/z* | |
|---|---|
| 31 | $CF^+$ |
| 69 | $CF_3{}^+$ |
| 100 | $CF_2CF_2{}^{\bullet+} (C_2F_4{}^{\bullet+})$ |
| 114 | $CF_2{=}N^+{=}CF_2 \ (C_2F_4N^+)$ |
| 119 | $CF_3CF_2{}^+ \ (C_2F_5{}^+)$ |
| 131 | $CF_2{=}CFCF_2{}^+ \ (C_3F_5{}^+)$ |
| 219 | $CF_3CF_2CF_2CF_2{}^+ \ (C_4F_9{}^+)$ |
| 264 | $CF_3CF_2CF_2CF{=}N^+{=}CF_2 \ (C_5F_{10}N^+)$ |
| 414 | $CF_3CF_2CF_2CF{=}N^+{=}CFCF_2CF_2CF_3 \ (C_8F_{16}N^+)$ |
| 464 | $CF_3CF_2CF_2CF_2{-}N^+{=}CF_2$ |
| | $\qquad\qquad\qquad\quad |$ |
| | $\qquad\qquad\qquad\ CF{=}CF{-}CF_2CF_3 \ (C_9F_{18}N^+)$ |
| 502 | $(CF_3CF_2CF_2CF_2)_2{-}N^+{=}CF_2 \ (C_9F_{20}N^+)$ |
| 576 | $CF_3CF_2CF_2CF_2{-}N^+{=}CF{-}CF{=}CF{-}CF_3$ |
| | $\qquad\qquad\qquad\quad |$ |
| | $\qquad\qquad\qquad\ CF{=}CF{-}CF_2CF_3 \ (C_{12}F_{22}N^+)$ |
| 614 | $(CF_3CF_2CF_2CF_2)_2{-}N^+{=}CF{-}CF{=}CF{-}CF_3 \ (C_{12}F_{24}N^+)$ |

**1.2.** 3-Ethylcyclohexene has the empirical formula $C_8H_{14}$, giving a molecular weight of 110, yet there is every reason to believe that the ion at $m/z$ 96 in Figure 1.22 represents the molecular ion, especially since it appears to lose 15 daltons (methyl radical) to give $m/z$ 81. In Chapter 4 we will also learn that organic compounds *cannot* lose 14 daltons $(110 - 96 = 14)$ from the molecular ion. Thus this spectrum undoubtedly is not that of 3-ethylcyclohexene!!

## CHAPTER 2

**2.1.** Assuming that $m/z$ 44 corresponds to the molecular ion, the sizes of $m/z$ 45 and 46 are consistent with those expected for isotope peaks of the $m/z$ 44 ion. (The possibility that this unknown does not produce a molecular ion, and that *all* of the observed ions are fragment ions, is discussed in Section 2.3.) There are only a limited number of organic compounds having a molecular weight of 44:

$$N_2O \quad CO_2 \quad CH_3CHO \quad (CH_2)_2O \quad CH_3CH_2CH_3 \quad FC{\equiv}CH$$
$$CH_3N{=}NH \quad CH_2NNH_2$$

The observed losses from the molecular ion are almost exclusively multiples of 12 and 16: 16 daltons to give the ion at $m/z$ 28, 28 daltons $(16 + 12)$ to give $m/z$ 16, 32 daltons $(16 + 16)$ to give the ion at $m/z$ 12. This fact, coupled with the lack of individual hydrogen losses (i.e., there are no peaks at $m/z$ 43, 42, 27, etc.), makes most of these possibilities seem unlikely. Eliminating those structures containing hydrogen, only two structures remain. Of these only $CO_2$ seems to fit all of the observed losses, since $N_2O$ would be expected to lose 14 daltons in additon to 16, not 12. Only the ion at $m/z$ 22 seems enigmatic, but it can be rationalized as the $M^{2+}$ ion, strengthening the assumption that $m/z$ 44 is the molecular ion. [*Answer:* Carbon dioxide; $CO_2$]

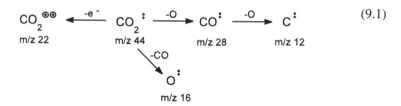

**2.2.**

$$\text{Average atomic weight of Br} = [(78.918)(50.52\%) + (80.916)$$
$$(49.48\%)]/100\%$$
$$= 79.906 \approx 80$$
$$\text{Average molecular weight of Br}_2 = 79.91 \times 2 = 159.82 \approx 160$$

These numbers have no meaning in mass spectrometry. The molecular ion region for $Br_2$ will show peaks at $m/z$ 158 for bromine molecules containing two atoms of $^{79}Br$, $m/z$ 160 for those containing one atom of $^{79}Br$ and one of $^{81}Br$, and $m/z$ 162 for those having two atoms of $^{81}Br$.

**2.3.** The probabilities for finding various combinations of chlorines in a molecule containing three chlorines are approximately as follows:

$$P(3\ ^{35}Cl) = (0.75)^3 = 0.422 \qquad P(2\ ^{35}Cl)(^{37}Cl) = (0.75)^2(0.25) = 0.141$$

$$P(^{35}Cl)(2\ ^{37}Cl) = (0.75)(0.25)^2 = 0.047 \qquad P(3\ ^{37}Cl) = (0.25)^3 = 0.016$$

When both isotopes are present at the same time, three different orientations are possible, so that

$$
\begin{aligned}
[M^+]/&[(M+2)^+]/[(M+4)^+]/[(M+6)^+] \\
&- P(3\ ^{35}Cl)/[3 \times P(2\ ^{35}Cl)(^{37}Cl)]/[3 \times P(^{35}Cl)(2\ ^{37}Cl)]/P(3\ ^{37}Cl) \\
&= (0.422)/[3 \times (0.141)]/[3 \times (0.047)]/(0.016) \\
&= 0.422/0.423/0.141/0.016 \\
&= 100/100/33/4
\end{aligned}
$$

Compare this answer with the graphic presentation in Figure 2.6.

**2.4.**

Size of $(M + 1)^+$ for $C_{60} = 60 \times 1.1\% = 66\%$, a large $(M + 1)^+$ ion!

Size of $(M + 2)^+$ for $C_{60} = (60 \times 1.1\%)^2/200 = 21.8\%$, a huge $(M + 2)^+$ ion!

With this many carbons, even the contributions of $(M + 3)^+$ and $(M + 4)^+$ (for three and four $^{13}C$'s, respectively) will be noticeable.

**2.5.** To facilitate determination of isotopic abundances, the relative intensities should be recalculated so that the intensity of the molecular ion is 100%. This is done by dividing each intensity by the relative abundance of the molecular ion (42.3%), giving

| $m/z$ | |
|---|---|
| 123 | 100.0% |
| 124 | 7.1% |
| 125 | 0.7% |

Two things should be apparent. First, the odd mass of the molecular ion dictates an odd number of nitrogens (Section 2.3). Second, the size of the $(M + 1)^+$ ion is inconsistent with more than six carbon atoms, since seven carbons would contribute $7 \times 1.1\% = 7.7\%$ to the $(M + 1)^+$ ion. In fact the intensity of the $(M + 1)^+$ ion accommodates exactly six carbons and one nitrogen very nicely: $(6 \times 1.1\%) + (1 \times 0.35\%) = 6.95\%$, well within experimental error of the observed intensity.

Six carbons and one nitrogen account for $(6 \times 12) + (1 \times 14) = 86$ daltons of the mass, so that 37 daltons remain to be assigned. Although the molecule could contain two more nitrogens (it cannot contain just one more because of the odd molecular weight), the $(M + 2)^+$ peak provides possible clues. Nitrogen contributes nothing to the $(M + 2)^+$ ion, but six carbons contribute $(6.6)^2\%/200 = 0.2\%$, well short of the observed intensity. The relatively small size of this ion dictates against some of the obvious choices (e.g., Si, S and Cl), and F would contribute nothing to this peak. The only choice remaining is oxygen. Two oxygens would contribute not only $0.2 \times 2 = 0.4\%$ to the $(M + 2)^+$ ion (giving a total of $0.2 + 0.4 = 0.6\%$; much closer to the observed intensity), but also 32 of the 37 missing units of mass.

The remaining 5 daltons are easily attributable to hydrogen, producing an empirical formula of $C_6H_5NO_2$, which is completely consistent with the observed data. Any attempt to draw a structure for this compound reveals that it must be highly unsaturated or contain several rings. From guideline 10 in Section 2.3, the number of rings plus double bonds in this molecule is $6 - \frac{1}{2}(5) + \frac{1}{2} + 1 = 6 - 2.5 + 0.5 + 1 = 5$. An aromatic compound would be a good starting place, and the peak at $m/z$ 77 (phenyl; $C_6H_5^+$) is consistent with that assumption. This accounts for three double bonds and one ring; the remaining double bond must be exterior to the ring. [*Answer:* Nitrobenzene; $C_6H_5NO_2$]

**2.6.** The presence of ion clusters at $m/z$ 83 to 87 and 118 to 122, each having major contributors separated by 2 daltons, strongly suggests the presence of chlorine and/or bromine. Comparison of peak intensities for the ions at $m/z$ 83, 85, and 87 with those in Figure 2.6 indicates that this ion probably contains two chlorines. If so, $m/z$ 118, being 35 daltons in mass above $m/z$ 83, is probably the molecular ion containing three chlorines. The relative intensities of the ions at $m/z$ 118, 120, and 122 are approximately 100:90:30, but because they are so weak, their intensities are known imprecisely. Again by comparison with Figure 2.6, this pattern most closely fits that for three chlorines.

Since the mass of three chlorines is 105 daltons, only 13 daltons remain unaccounted for. The only reasonable possibility is one carbon and one hydrogen, giving $CHCl_3$ as the empirical formula. This is confirmed by the abundance of the ion at $m/z$ 84 (1.3%), which is consistent with the presence of one carbon in $m/z$ 83. Since 83 daltons is 13 more than the combined mass of two chlorines, its structure must be $CHCl_2^+$.

The remainder of the spectrum is consistent with this assignment. The cluster at $m/z$ 70 appears to contain two chlorines ($Cl_2^+$), while those following $m/z$ 47 and 35 are more complex. These clusters make sense if we recognize that each contains two different ions whose isotope clusters overlap. Thus $m/z$ 48 is due to $CHCl^+$ (with its $^{37}Cl$ isotope peak at $m/z$ 50), while $m/z$ 47 comes from $CCl^+$ (with its corresponding isotope peak at $m/z$ 49). The ions at $m/z$ 36 and 35 (with isotope peaks at $m/z$ 38 and 37, respectively) correspond to $HCl^+$ and $Cl^+$. [*Answer:* Chloroform; $CHCl_3$]

(9.2)

$$CI_2^{\cdot\,:} \xleftarrow{-CHCI} CHCl_3^{\cdot\,:} \xrightarrow{-CI\,\cdot} CHCl_2^{\oplus} \xrightarrow{-CI\,\cdot} CHCl^{\cdot\,:} \xrightarrow{-H\,\cdot} CCl^{\oplus}$$

m/z 70     m/z 118     m/z 83     m/z 48     m/z 47

$$\downarrow -CCl_2$$

$$HCl^{\cdot\,:} \xrightarrow{-H\,\cdot} CI^{\oplus}$$

m/z 36     m/z 35

**2.7.** The molecular ion at $m/z$ 76 has two isotope peaks that should draw attention. The $(M + 2)^+$ ion is more intense than the $(M + 1)^+$, yet not large enough to be due to chlorine or bromine. Silicon also can be ruled out since the isotopic contribution from $^{29}Si$ is greater than that of $^{30}Si$, leaving only oxygen and sulfur to be considered. The $(M + 2)^+$ contribution for oxygen is much too small (0.2% per atom; Table 2.1) to account for the relatively large peak observed here. Thus sulfur must be the heteroatom.

Each sulfur atom contributes $0.76/95.0 = 0.8\%$ to the $(M + 1)^+$ ion and $4.2/95.0 = 4.4\%$ to the size of the $(M + 2)^+$ ion. The observed intensity of $(M + 2)^+$ in this spectrum is 8.5%, much too large to be consistent with one sulfur even if minor contributions from other common elements are added to it. On the other hand, two sulfurs contribute $2 \times 4.4\% = 8.8\%$ to the $(M + 2)^+$ peak, within experimental error of the observed intensity. The presence of two sulfurs accounts for only 64 daltons of mass, 12 short of the molecular weight of 76. This difference is filled by a single carbon atom, leading to $CS_2$ as the likely answer.

This structure fits the remaining data in the spectrum. First, the contributions of two sulfurs to the $m/z$ 77 ion should be $2 \times 0.8\% = 1.6\%$, with one carbon atom contributing an additional 1.1%. This total of 2.7% is almost exactly that observed. Next, apparent losses from the molecular ion are of 32 daltons to give the $m/z$ 44 ion, 38 daltons to give the $m/z$ 38 ion, and of 44 daltons $(32 + 12)$ to give the $m/z$ 32 ion. In addition, the ions at $m/z$ 32 and 44 each appear to contain exactly one S, based on the relative intensities of each of the corresponding $(P+2)^+$ ions. As with $CO_2$, the $m/z$ 38 ion is not a fragment ion, but rather the $M^{2+}$ ion. [*Answer:* Carbon disulfide; $CS_2$]

$$CS_2^{\oplus\oplus} \xleftarrow{-e^-} CS_2^{\cdot\,:} \xrightarrow{-S} CS^{\cdot\,:}$$

m/z 38     m/z 76     m/z 44

(9.3)

$$\downarrow -CS$$

$$S^{\cdot\,:}$$

m/z 32

**2.8.** The molecular weight of this compound appears to be even, making the presence of nitrogen unlikely. The abundances of both the $(M + 1)^+$ ion at $m/z$ 67

(1.1% relative to $m/z$ 66) and $m/z$ 48 [$(P+1)^+$ for the base peak] are consistent with, at most, the presence of one carbon. This leaves $66 - 12 = 54$ daltons of mass to account for, and there are no further clues from the isotopic abundances associated with either ion about what other elements might be present. On the other hand, the *lack* of clues in itself is a clue, and may indicate the presence of atoms that show no isotopic abundance pattern.

The major loss from the molecular ion is 19 daltons ($66 - 47$). The most likely candidate to account for this loss is fluorine, an atom having one naturally occurring isotope. The other major ion in the spectrum occurs at $m/z$ 28, 19 daltons below $m/z$ 47, strong evidence that there may be more than one fluorine atom in the molecule. Two fluorines and one carbon have a mass of 50 daltons, leaving only 16 daltons unascribed. The small ion at $m/z$ 50, resulting from loss of 16 daltons from the molecular ion, might indicate the presence of an $NH_2$ group, but because of the odd-nitrogen rule the presence of one nitrogen would necessitate a second nitrogen due to the even molecular weight. The presence of two nitrogen atoms is thus inconsistent with both the isotopic abundance data and the fragmentation pattern, leaving oxygen as the most likely candidate for the remaining mass. All of the fragmentations of this molecule are consistent with this assignment. [*Answer:* Carbonyl difluoride; FCOF)

$$CO^{+\bullet} \xleftarrow{-F\bullet} COF^{\oplus} \xleftarrow{-F\bullet} COF_2^{+\bullet} \xrightarrow{-\ddot{\overset{\bullet\bullet}{O}}\!:} CF_2^{+\bullet} \xrightarrow{-F\bullet} CF^{\oplus} \quad (9.4)$$

$m/z$ 28     $m/z$ 47     $m/z$ 66     $m/z$ 50     $m/z$ 31

$\downarrow$ -CO         $\downarrow$ -CF$_2$

$F^{\oplus}$           $O^{+\bullet}$

$m/z$ 19        $m/z$ 16

**2.9.** This spectrum, like that in Example 2.1, exhibits both an odd molecular weight (157) and a molecular ion cluster having a doublet of equally intense peaks separated by 2 daltons. Because of the odd-nitrogen rule, this compound must contain at least one nitrogen. The intense doublet at $m/z$ 157 and 159 is consistent with the presence of one bromine, which is lost ($157 - 79$) to produce the single ion at $m/z$ 78.

The ions at $m/z$ 158 and 160 are the "$(M + 1)^+$" ions for the peaks corresponding to the two isotopes of bromine; both provide the same information concerning the number of carbon and nitrogen atoms in the compound (in fact, so does $m/z$ 79 since there is no ion corresponding to $^{81}Br^+$). If we assume the presence of only one nitrogen, the $^{13}C$ contribution to $m/z$ 158 is $6.0 - 0.4 = 5.6\%$, consistent with the presence of five carbons. Five carbons, a nitrogen, and a bromine add up to $(5 \times 12) + (1 \times 14) + (1 \times 79) = 153$ daltons, leaving 4 daltons to be accounted for by hydrogen.

The empirical formula $C_5H_4NBr$ gives rise to $5 - \frac{1}{2}(5) + \frac{1}{2} + 1 = 4$ rings plus double bonds—a highly unsaturated molecule (note that the bromine counts as a hydrogen in this equation—see Section 2.3). A simple structure that fulfills these requirements is bromopyridine. Three isomeric structures are possible, and we cannot distinguish between them without obtaining standard spectra for each one. [*Answer:* 3-Bromopyridine]

(9.5)

         m/z 157                 m/z 78               m/z 51

**2.10.** At first glance, this spectrum provides few clues about its solution. The apparent molecular ion ($m/z$ 146) is even, and obvious halogen isotope patterns are missing from all of the major ion clusters. If we assume the presence of only carbon, hydrogen, and oxygen, the $(M + 1)^+$ ion at $m/z$ 147, which is $1.8/10.4 = 16.8\%$ relative to the molecular ion, indicates that there should be 15 carbons in the molecule. This is impossible with a molecular weight of 146! Furthermore, the $(M + 2)^+$ ion, at about 7% of the molecular ion, is much too large to accommodate any reasonable number of carbons and oxygens. The same observations hold for the ion clusters at $m/z$ 131 to 133 and 73 to 75.

Although the presence of one sulfur could be invoked to help explain part of the relative abundances of $m/z$ 133 and 148, both of these ions are still too large for any combination of carbon, oxygen, and one sulfur. Two sulfurs, on the other hand, would lead to an $(M + 2)^+$ ion having a relative intensity of about 8.8% (see Problem 2.6), far outside the range of experimental error for these peaks.

This situation is similar to that encountered in Example 2.3; what we need to solve this problem is an element that has sizable contributions to both the $(M + 1)^+$ and $(M + 2)^+$ ions, but with the contribution to the $(M + 2)^+$ smaller than that shown by sulfur. As in Example 2.3, the answer is silicon. This choice looks even more attractive when we consider that the base peak in the spectrum is $m/z$ 73, an ion commonly observed in the spectra of compounds containing trimethylsilyl groups (Section 2.2.1.6). Let us continue, then, under the assumption that there is at least one silicon in the molecule.

One silicon contributes 5.1% to the $(M + 1)^+$ ion and 3.5% to the $(M + 2)^+$ ion (see Example 2.3). Clearly the intensities of the ions at $m/z$ 133 and 148 (both at about 7% or more abundance relative to $m/z$ 131 and 146, respectively) make the presence of two silicons in both of these ions seem likely. Under this assumption, let us determine the remaining carbon, hydrogen, and oxygen content in these ions.

For $m/z$ 147, $2 \times 5.1\% = 10.2\%$ of the relative abundance is accounted for by two silicons, leaving $16.8 - 10.2 = 6.6\%$ for contributions from $6.6\%/1.1\% = 6$ carbons. Two silicons also account for essentially all of the $m/z$ 148 ion ($2 \times 3.5\% = 7.0\%$), with a minimal addition due to two $^{13}$C's. The $m/z$ 131 ion cluster similarly shows the presence of two silicons and five carbons (one carbon atom having been lost as methyl radical—15 daltons):

$$m/z\ 132 \Rightarrow 2 \times 5.1\%\ (^{29}Si) + 5 \times 1.1\%\ (^{13}C) = 15.7\%\ (3.3/20.9$$
$$= 15.8\%\ obs.)$$

$$m/z\ 133 \Rightarrow 2 \times 3.5\%\ (^{30}Si) + [(5 \times 1.1\%)^2/200\ (^{13}C)]$$
$$= 7.2\%\ (1.5/20.9 = 7.1\%\ obs.)$$

Two silicons and six carbons lead to $(2 \times 28) + (6 \times 12) = 128$ daltons of mass, leaving 18 daltons unaccounted for. Although two hydrogens and an oxygen offer a possible explanation, 18 hydrogens seems more likely, especially if $m/z$ 73 is indeed due to the trimethylsilyl ion. A probable arrangement of these atoms is $(CH_3)_3Si-Si(CH_3)_3$, although other isomeric structures cannot be ruled out at this point. [*Answer:* Hexamethyldisilane]

$$(9.6)$$

**2.11.** The molecular weight for this compound is even, and no halogens are present. The $(P+2)^+$ ions for both $m/z$ 94 and 79 are larger than their respective $(P+1)^+$ ions, and of the right relative abundance to indicate the presence of one sulfur (see Section 2.2.2.1 and Figure 2.11). At least one carbon is indicated by the fact that the molecular ion loses 15 daltons (methyl radical) to give the base peak at $m/z$ 79.

The relative abundances in the molecular ion cluster are $1.8/57.3 = 3.1\%$ for the $(M + 1)^+$ ion and $2.8/57.3 = 4.9\%$ for the $(M + 2)^+$ ion. One sulfur contributes 0.8% to the $(M + 1)^+$ ion, leaving $3.1 - 0.8 = 2.3\%$ to be accounted for by any remaining elements. Since nitrogen is not indicated and at least one carbon is likely, two carbons ($2 \times 1.1\%$ per carbon) seem like a reasonable solution. For the $(M + 2)^+$ ion, one sulfur accounts for 4.4%, leaving about 0.5% for other contributors. This value is much too large for two $^{13}C$'s, but is just about right for two oxygen atoms.

An analysis of the isotope cluster at $m/z$ 79 is consistent with this interpretation:

$m/z$ 80:     1.9% $\approx$ (1 $\times$ 0.8%; $^{33}$S) + (1 $\times$ 1.1%; $^{13}$C)

$m/z$ 81:     4.9% $\approx$ (1 $\times$ 4.4%; $^{34}$S) + (2 $\times$ 0.2%; $^{18}$O)

Two carbons, two oxygens, and a sulfur account for (2 $\times$ 12) + (2 $\times$ 16) + (1 $\times$ 32) = 88 daltons, leaving six hydrogens to make up the remaining mass. Several different arrangements of these atoms are possible that might produce a spectrum such as this, and standard spectra of each would have to be run to ensure complete identification. [*Answer:* Dimethylsulfone; $CH_3SO_2CH_3$]

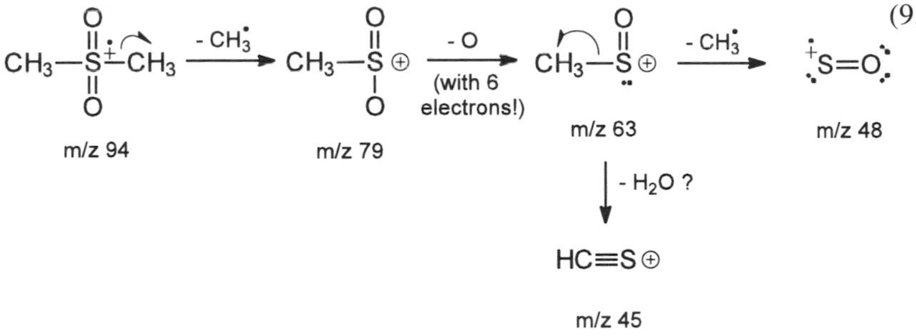

(9.7)

# CHAPTER 3

**3.1.** Hydrogen and methyl ions are much less stable then benzyl; thus, reactions leading to their formation provide little stabilization to the corresponding transition states. Also, the fragmentations producing these ions already have an additional energy barrier due to the charge migration mechanism.

**3.2.**

a.

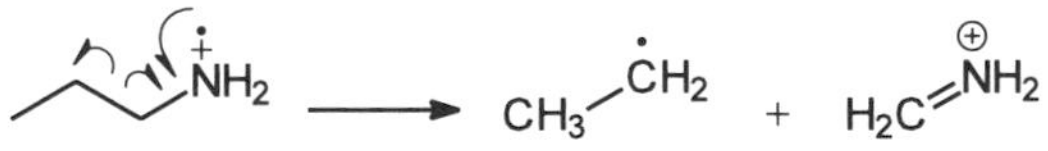

(9.8)

b.

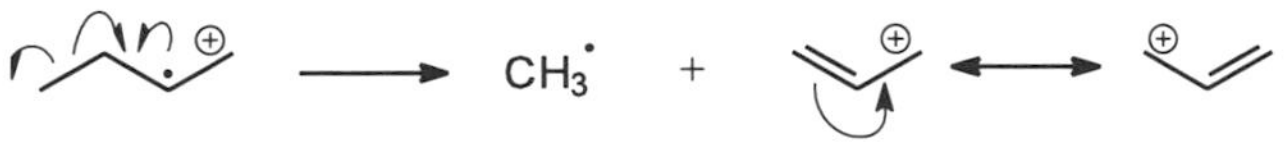

(9.9)

c.

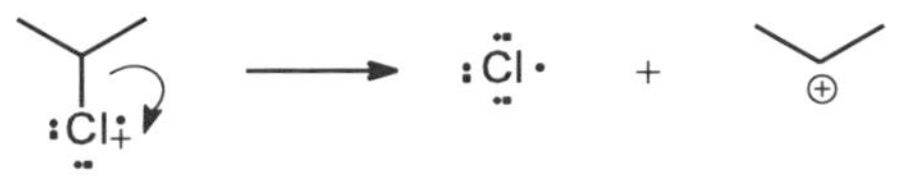

(9.10)

d.

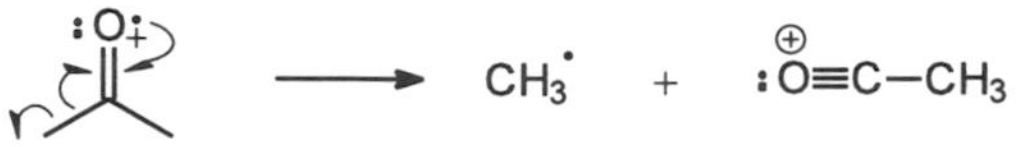

(9.11)

e.

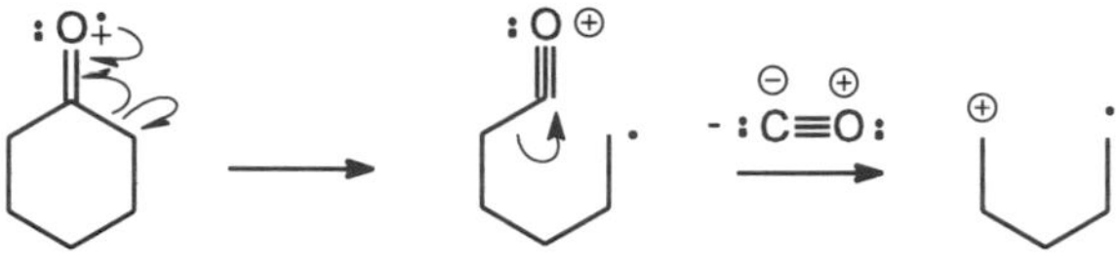

(9.12)

## CHAPTER 4

**4.1.** Even without examining the fragmentations of this compound in detail, a quick perusal of the spectrum reveals the following:

a. The molecular weight is even (112), so that nitrogen is probably absent.
b. The molecular ion isotope cluster shows an $(M + 1)^+$ ion of 6.7% and an $(M + 2)^+$ ion of 32.7%, consistent with the presence of six carbon atoms and one chlorine atom, respectively.
c. An aromatic low mass ion series occurs at $m/z$ 38, 51, and 77, indicating an aromatic ring with an attached electron-withdrawing group (in this case chlorine).
d. The only major loss from the molecular ion is that of chlorine radical $(112 - 77 = 35$ daltons), confirmed by the absence of a $^{37}Cl$ isotope peak at $m/z$ 79.

Six carbons and a chlorine make up $(6 \times 12) + (1 \times 35) = 107$ daltons of the mass. The remainder must come from five hydrogen atoms, giving the empirical formula $C_6H_5Cl$. The small peak at $m/z$ 97 corresponds to the loss of methyl radical, as in benzene, and the pair of peaks at $m/z$ 56 and 57, which in this case occur at *adjacent* masses in a ratio of 3:1, must be due to the $M^{2+}$ ion. [*Answer:* Chlorobenzene; $C_6H_5Cl$]

(9.13)

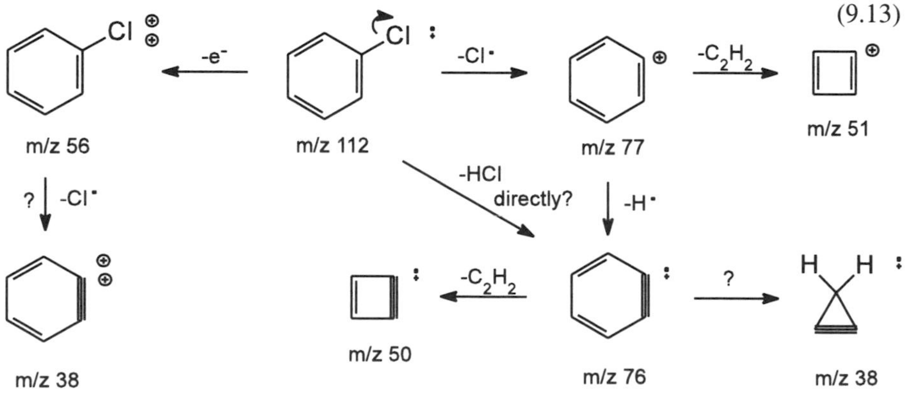

**4.2.** There are nine isomeric $C_7H_{16}$ structures:

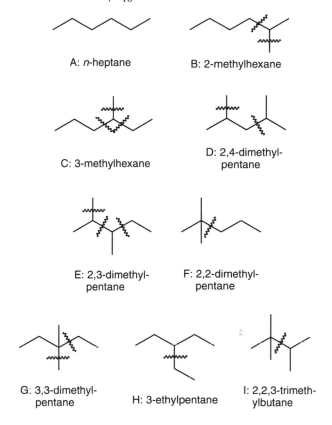

A: *n*-heptane    B: 2-methylhexane

C: 3-methylhexane    D: 2,4-dimethyl-pentane

E: 2,3-dimethyl-pentane    F: 2,2-dimethyl-pentane

G: 3,3-dimethyl-pentane    H: 3-ethylpentane    I: 2,2,3-trimeth-ylbutane

Based on the discussions in this chapter, these structures lead us to expect the following spectral features:

A: A "typical" *n*-alkane spectrum, with $(M-15)^+$ being the smallest of the fragment ions in the alkane ion series. Spectrum b in Figure 4.14 is a possibility for this structure.

B: Loss of either methyl or butyl radical generates a secondary carbocation, so that ions at $m/z$ 85 and 43 should be prominent. Loss of an ethyl group is not expected, since this generates a primary carbocation and a primary radical. Loss of isopropyl radical to give $m/z$ 57, however, may be significant due solely to the relative stability of the radical. Spectrum c in Figure 4.14 is consistent with these expectations.

C: Loss of methyl, ethyl, or propyl radical leads to a secondary ion, thus ions at $m/z$ 85, 71, and especially $m/z$ 57 should be intense. Although spectrum b (Figure 4.14) is a possibility, we might expect $m/z$ 85 for this compound to be larger than the molecular ion (compare the spectrum of 2-methylheptane; Figure 4.13*b*).

D: The options open to this molecule are similar to those of B. Predicting differences between the spectra of these two compounds is not meaningful at this point.

E: This structure, like C, can form secondary ions by loss of methyl, ethyl, or isopropyl. Because propyl loss occurs as isopropyl radical in this case, formation of an intense $m/z$ 57 ion (relative to $m/z$ 71 and 85) is even more likely. Again, although spectrum b (Figure 4.14) is a possibility, the molecular ion seems too large and the $m/z$ 57 ion too weak in intensity to fit this structure.

F: Because loss of $n$-propyl radical leads to the $t$-butyl carbocation, $m/z$ 57 should be the base peak in the spectrum [see Figure 4.13$c$ and Eq. (4.6)]. This is not observed in any of the spectra.

G: Loss of either a methyl or an ethyl group gives a tertiary ion, so that peaks at both $m/z$ 85 and 71 should be prominent features of the spectrum. No loss of propyl is expected. Spectrum a (Figure 4.14) is consistent with this structure.

H: Loss of hydrogen radical from the central carbon leads to a tertiary ion, while loss of any of the three ethyl groups should produce a strong $m/z$ 71 ion. These features are not observed in any of the spectra.

I: Like F, the stability of both the $t$-butyl carbocation and the isopropyl radical should produce a spectrum with $m/z$ 57 as the base peak. This is not observed in any of the spectra.

Structures F, H, and I appear to be inconsistent with the observed spectra in Figure 4.14 and can be eliminated almost immediately. Although structures C and E are consistent with spectrum b, neither seems like the best candidate. On the other hand, structure G seems consistent with spectrum a, structure A with spectrum b, and both structures B and D with spectrum c. In fact, this is as far as we can go on the basis of the available data. Spectrum c actually belongs to structure D, but without the spectrum of B for comparison, this assignment is uncertain.

Spectra of the remaining compounds are shown in Figure 9.1 for comparison. Predictably, there are some surprises. For example, $m/z$ 85 in the spectrum of compound C is smaller than expected, although it is still not smaller than the molecular ion. Also, compounds E and H both show intense rearrangement ions at $m/z$ 56 and 70, respectively, that compete in relative abundance with the predicted fragment ions. This tendency is more pronounced in compounds that can form highly substituted olefins (see also discussions of the $\gamma$-hydrogen rearrangement in Chapter 6). [*Answer:* (a) 3,3-Dimethylpentane; (b) $n$-heptane; and (c) 2,4-dimethylpentane]

**4.3.** An overview of the spectrum produces the following information:

a. The even molecular weight mitigates against the presence of nitrogen in the molecule.

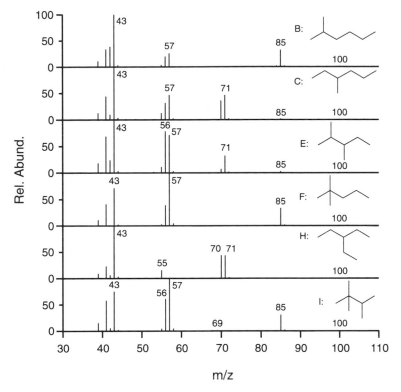

**Figure 9.1.** Mass spectra of the $C_7H_{16}$ isomers not included in Problem 4.2.

b. The ions at $m/z$ 39 and 50, the intense molecular ion, and the major loss of acetylene (84 − 58) are indicative of an aromatic compound, although obviously not a benzene derivative (phenyl itself has a mass of 77 daltons).

c. The $(M + 2)^+$ ion seems too large for compounds containing only carbon, hydrogen, and oxygen.

Although the relative sizes of $m/z$ 85 and 86 could be caused by either sulfur or silicon, the isotope cluster associated with $m/z$ 58 is more consistent with sulfur ($m/z$ 60, about 5% relative to $m/z$ 58, cannot be a fragment ion since it is 24 daltons less than the molecular ion).

If the molecular ion contains one sulfur, the $(M + 1)^+$ ion must have a 0.8% contribution from $^{33}S$ and the $(M + 2)^+$ ion a 4.4% contribution from $^{34}S$. The intensity of the $(M + 2)^+$ ion looks about right for one sulfur, but the intensity of the $(M + 1)^+$ ion obviously reflects the presence of other elements. A good first guess is carbon (since the loss of acetylene is suspected already, and the $m/z$ 39 ion is probably the cyclopropenium ion). Subtracting 0.8% (for $^{33}S$) from the observed intensity of the $(M + 1)^+$ ion gives a 5.3 − 0.8 = 4.5% contribution from carbon, consistent with the presence of four carbon atoms in the molecule.

The total mass of one sulfur and four carbons is $32 + (4 \times 12) = 80$ daltons. The remaining 4 daltons must be due to hydrogens.

The empirical formula thus appears to be $C_4H_4S$, and, despite the loss of a methyl group, the compound appears to be aromatic from the low mass ion series. The problem is one of drawing an aromatic structure for this formula. The number of rings plus double bonds (Section 2.3) is $4 - \frac{1}{2}(4) + 0 + 1 = 3$. The answer is thiophene, a five-membered, unsaturated, sulfur-containing ring, which, in addition to two nonbonding electrons donated by the sulfur, has four π-electrons from the two double bonds to form an aromatic ring. This mass spectrum, in fact, offers support for the aromaticity of the thiophene ring. Loss of methyl from this compound is as enigmatic as it is from benzene. [*Answer:* $C_4H_4S$; thiophene]

(9.14)

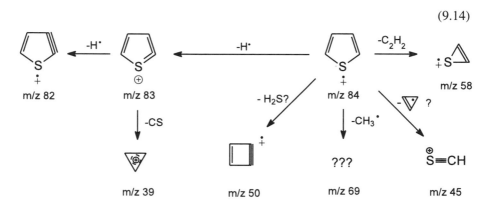

**4.4.** The low mass ion series should lead quite rapidly to the conclusion that this compound is a saturated alkane. The molecular ion appears to be at $m/z$ 128, and the isotopic abundance data are consistent with nine carbons ($m/z$ 129 is $10\% \approx 9 \times 1.1\%$ relative to $m/z$ 128). Determining a unique structure for this compound is more difficult, especially without other spectra for comparison. Despite the fact that $m/z$ 85 is larger than $m/z$ 71, this is a real-life spectrum of the straight-chain isomer. [*Answer:* $C_9H_{20}$; *n*-nonane]

**4.5.** It is instructive to compare this spectrum with that of the unknown in Problem 4.1. The two spectra are similar in overall appearance, but the masses of the two most intense ions shift upward by 1 dalton in the spectrum of this unknown.

From first impressions, the following can be deduced:

a. The compound has an odd molecular weight, so must contain at least one nitrogen.

b. The ions at $m/z$ 113 and 86 each contain one chlorine, and the ion at $m/z$ 78 reflects loss of the chlorine.

c. The low mass ion series at $m/z$ 51 and 78 and the intense molecular ion are consistent with an aromatic structure.

d.  The first loss from the molecular ion is 27 daltons (HCN), indicating that the nitrogen is either directly attached to, or is in, the aromatic ring.

e.  The isotopic abundance data for the molecular ion is consistent with five carbons and one nitrogen [$(M + 1)^+$: 5.5% for 5 carbons + 0.4% for N = 5.9%].

This accounts for $(5C \times 12) + (1N \times 14) + (1Cl \times 35) = 109$ daltons. The remaining 4 mass units must come from hydrogen. An aromatic structure that accommodates this empirical formula is chloropyridine. Three isomeric structures are possible, but they are distinguishable only by comparing the spectra of actual standards. [*Answer:* $C_5H_4NCl$; 3-chloropyridine]

(9.15)

m/z 86          m/z 113          m/z 78          m/z 51

**4.6.** The presence of a single, intense fragment ion at low mass is a strong signal that a heteroatom-containing functional group may be present in an aliphatic portion of the molecule. The specific type of functional group can be narrowed down by reference to the list of low mass ion series in Table 4.2. In this case, the large ion at $m/z$ 45 fits into the series for aliphatic ethers and alcohols. The isotopic abundance of the $m/z$ 46 ion indicates the presence of two carbons, consistent with $C_2H_5O$. The ion at $m/z$ 74 appears to be the molecular ion, since the first observed loss is 15 daltons to give $m/z$ 59. The major loss from $m/z$ 74 to 45 is 29 daltons, which could be either an ethyl group or HCO.

The isotopic abundance seen in $m/z$ 75 is consistent with four carbons, although caution is advised since this peak is so small (if the intensity of $m/z$ 75 is actually 0.7, rather than 0.8, the relative percentage is 3.9, not 4.5%). From the peaks at $m/z$ 45 and 74, then, a molecular formula of $C_4H_{10}O$ seems likely. We cannot proceed further without additional knowledge about how compounds such as this fragment. We will return to this problem in Chapter 5 and determine the actual structure. [*Answer:* $C_4H_{10}O$; the answer is *not* diethylether—see Chapter 5]

**4.7.** Although this compound has a fairly high molecular weight, thus increasing the difficulty of unambiguous identification, a number of features about this spectrum should narrow down the list of possibilities:

a.  The molecular ion, at even mass, is over four times larger than any other ion in the spectrum.

b.  There is a distinctive electron-withdrawing aromatic low mass ion series.

c.  Other than hydrogen, the first loss from the molecular ion is 26 daltons (probably due to acetylene, since nitrogen is not indicated) to give the tiny peak at $m/z$ 128, and $m/z$ 128 appears to lose 26 daltons to produce $m/z$ 102.

This all points to a highly aromatic compound—probably one with more than one aromatic ring.

The isotopic abundance data from the molecular ion cluster indicates the presence of 12 carbons ($12 \times 1.1\% = 13.2\%$). If this is reliable, carbon alone accounts for 144 of the 154 daltons of mass, so the remaining 10 daltons must come from hydrogen. Although several highly unsaturated $C_{12}H_{10}$ isomers can be drawn [rings plus double bonds = $12 - \frac{1}{2}(10) + 0 + 1$], the most common is biphenyl. In reality the spectra of several of these compounds are so similar to this one that alternative answers (acenaphthalene, e.g.) are acceptable. [*Answer:* $C_{12}H_{10}$; biphenyl]

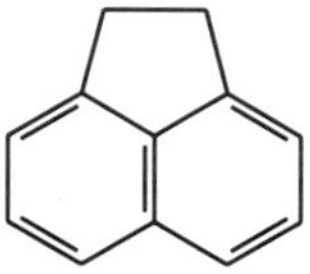

Acenaphthalene

(9.16)

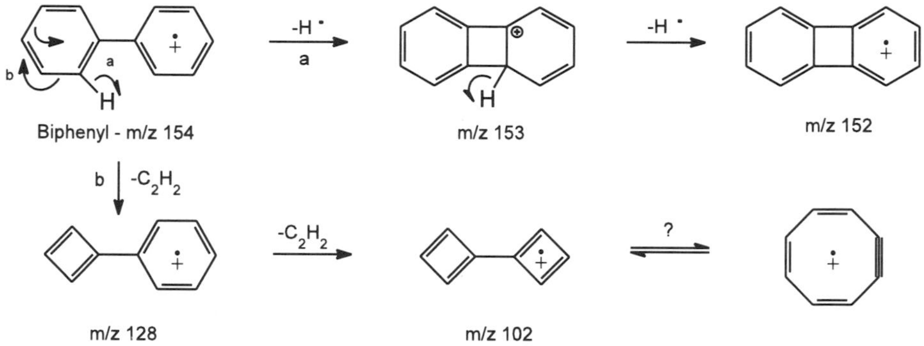

Biphenyl - m/z 154      m/z 153      m/z 152

m/z 128      m/z 102

## CHAPTER 5

**5.1.** These compounds have the same, even, molecular weight and appear to be isomeric with one another. From the isotopic abundance data in the molecular ion clusters, we find that both compounds appear to have nine carbons. Although the intensity of $m/z$ 92 in Figure 5.6$a$ indicates the presence of 9 to 10 carbons for $m/z$ 91 in this spectrum, this is impossible for the mass of the ion. This means only that another ion is contributing to the size of the $m/z$ 92 ion (see Chapter 6). Since the presence of hydrogen is suspected from both spectra, the remaining 12 daltons in each case are probably contributed by hydrogen.

Both spectra show an aromatic low mass ion series, which is not surprising given the empirical formulas. Four basic types of structures are possible for aromatic $C_9H_{12}$ isomers:

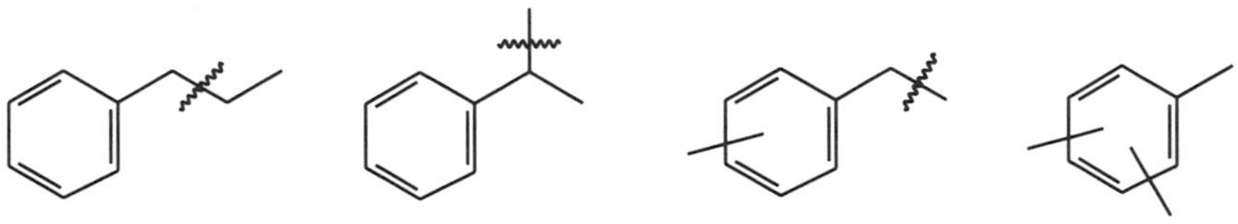

Of these, only *n*-propylbenzene is expected to lose an ethyl group by benzylic cleavage. This loss produces an abundant $(M - 29)^+$ ion at *m/z* 91, like that seen in Figure 5.6*a*.

Isopropylbenzene and the isomeric methylethylbenzenes, on the other hand, should each lose a methyl group by benzylic cleavage, thereby producing an abundant *m/z* 105 ion like that in Figure 5.6*b*. The final group of compounds, the isomeric trimethylbenzenes, should lose neither methyl nor ethyl groups by this process, although 1,2,3-trimethylbenzene loses the central methyl group because of steric strain (Figure 1.15*c*). In any case, all of these compounds should produce an ion of at least modest intensity at *m/z* 119 due to loss of hydrogen by benzylic cleavage. The choice between isopropylbenzene and the methylethylbenzenes depends on more subtle criteria. Isopropylbenzene has only one hydrogen available for loss by benzylic cleavage, whereas the methylethylbenzenes all have five. The latter compounds thus should exhibit at least a weak *m/z* 119 ion (compare with Figures 1.15 and 5.3). Since Figure 5.6*b* shows no ion visible at *m/z* 119, isopropylbenzene seems like a more reasonable choice. [*Answer:* (a) *n*-Propylbenzene and (b) isopropylbenzene]

**5.2.** The molecular ion cluster around *m/z* 188 in the spectrum of CS is consistent with the presence of one chlorine. The losses from the molecular ion can be rationalized as follows:

*m/z*
161    $(-HCN)$
153    ($-Cl\cdot$; confirmed by isotopic abundances)
137    ($-HCCCN$, cyanoacetylene; loss of acetylenes from aromatic compounds is typical, although loss of this particular acetylene seems unusual)
126    ($-Cl\cdot$, $-HCN$)
100    ($-Cl\cdot$, $-HCN$, $-HCCH$)
99     ($-Cl\cdot$, $-HCN$, $-HCN$)

An aromatic low mass ion series occurs around the ions at *m/z* 39, 50, 62, and 75.

The molecular ion of the unknown occurs at $m/z$ 190 (2 daltons higher than CS) and contains one chlorine. Since it seems reasonable to assume that this compound might be structurally related to CS, a logical choice for a structure is one in which hydrogen has been added across the C—C double bond of CS. In contrast to CS, which is highly unsaturated and fragments by losing chlorine radical and small unsaturated molecules, this compound has a fragile bond that can undergo α-cleavage (shown here after initial ionization at chlorine) to produce the intense ion at $m/z$ 125:

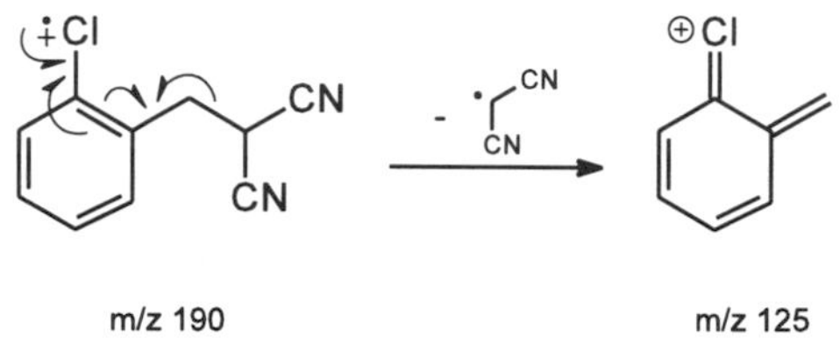

(9.17)

m/z 190        m/z 125

The remaining fragmentations of the unknown are unexceptional. Loss of HCN gives $m/z$ 163, while loss of Cl· yields the tiny peak at $m/z$ 155. These two compounds illustrate well the difference in behavior between aromatic and "aliphatic" compounds of similar structure. Notice that solving this problem, involving a compound with a fairly complex molecular structure, was made easier because we knew something about the "chemical history" of the sample (Section 2.3, guideline 2). [*Answer:* 1-(2-Chlorophenyl)-2,2-dicyanoethane]

**5.3.** (i) The presence of an important $m/z$ 91 ion, as opposed to $m/z$ 105 or 119, indicates that the aromatic ring and its attached carbon contain no methyl groups. (ii) The base peak at $m/z$ 58 limits the distribution of alkyl groups in the vicinity of the nitrogen atom and thus the number of structures that need to be considered as possible solutions. In fact, only five structures, shown in Table 9.1 with their predicted α-cleavage losses, are possible.
(iii) The spectrum in Figure 5.12e is that of methamphetamine. Comparison of Figures 5.10a and 5.12d reveals that the latter spectrum is that of phentermine.

These structures can be categorized fairly easily according to their respective α-cleavage losses: A and E—hydrogen, methyl, and benzyl radicals; B—methyl and benzyl radicals; C—hydrogen, ethyl, and benzyl radicals; and D—only hydrogen and benzyl radicals. Since Figure 5.12b shows no significant losses of either methyl or ethyl groups, it seems most suitable for structure D. Likewise, Figure 5.12c exhibits a relatively intense peak for loss of an ethyl group, consistent with C, and Figure 5.12d shows loss of methyl, but not hydrogen, as the spectrum for B should (and does; see Figure 5.10b).

Assigning structures to spectra in Figure 5.12a and e would be nearly impossible without additional information. By comparison with a known spectrum, we know that Figure 5.12e is the spectrum of methamphetamine (A in Table 9.1). This is consistent with the difference between structures A and E. The peaks at $m/z$ 115 and 117, which for methamphetamine come from the aromatic ring and its attached

**Table 9.1. Possible Solutions for Problem 5.3**

| Structure | α-Cleavage Losses |
|---|---|
| A. | 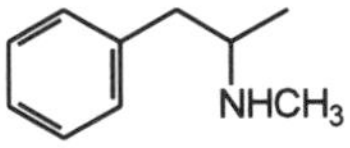 4 H$^{•}$ ($m/z$ 148); CH$_3^{•}$ ($m/z$ 134); φCH$_2^{•}$ ($m/z$ 58) |
| B. | 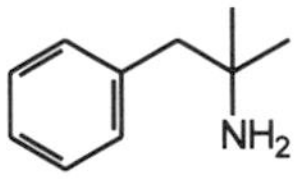 2 CH$_3^{•}$ ($m/z$ 134); φCH$_2^{•}$ ($m/z$ 58) |
| C. | 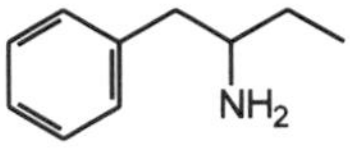 H$^{•}$ ($m/z$ 148); CH$_3$CH$_2^{•}$ ($m/z$ 120); φCH$_2^{•}$ ($m/z$ 58) |
| D. | 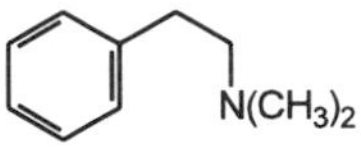 8 H$^{•}$ ($m/z$ 148); φCH$_2^{•}$ ($m/z$ 58) |
| E. | 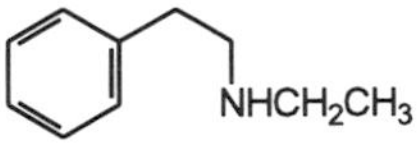 5 H$^{•}$ ($m/z$ 148); CH$_3^{•}$ ($m/z$ 134); φCH$_2^{•}$ ($m/z$ 58) |

3-carbon chain [see Eq. (7.5)], would not be expected in the spectrum of structure E, which has only a 2-carbon side chain and thus shows anenhanced peak at $m/z$ 105. [*Answer:* (a) *N*-ethyl-β-phenethylamine; (b) *N,N*-dimethyl-β-phenethylamine; (c) 1-phenyl-2-aminobutane; (d) phentermine; and (e) methamphetamine]

**5.4.** Although benzyl ion ($m/z$ 91) can be formed by charge migration α-cleavage from this compound, ions in which the positive charge is stabilized by nitrogen form more readily (see, e.g., Problem 5.3). Thus, only ions of the latter type are listed below:

a. $m/z$ 295 (loss of H·):

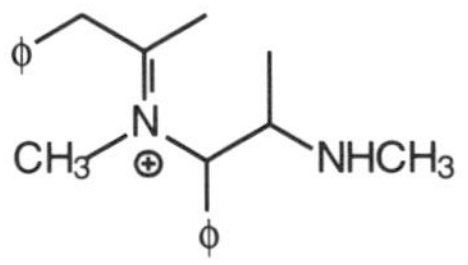

A 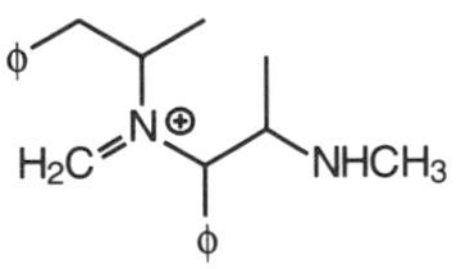          B 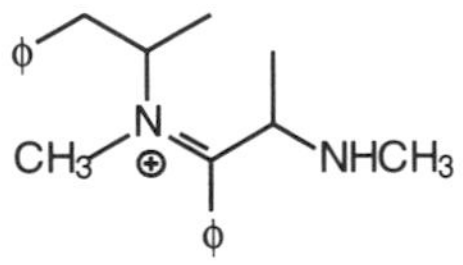          C

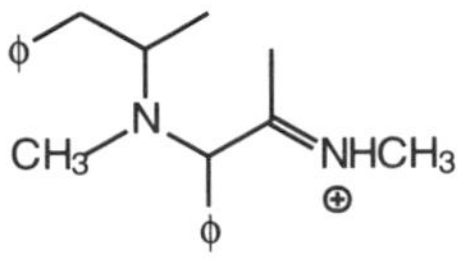

D 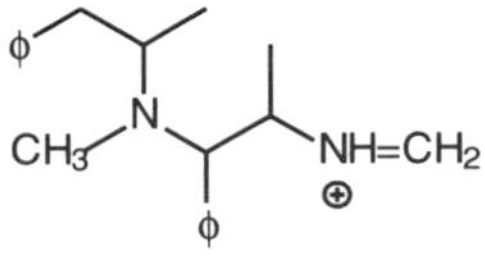          E

b. *m/z* 281 (loss of CH$_3$·):

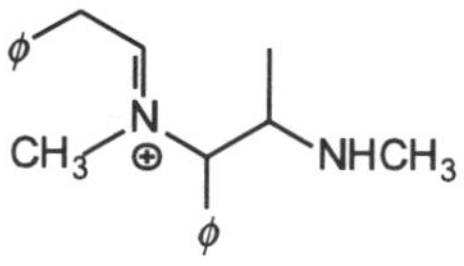

F

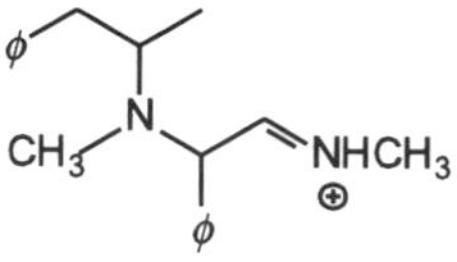

G

c. *m/z* 238:

H

d. *m/z* 219 (loss of φ·):

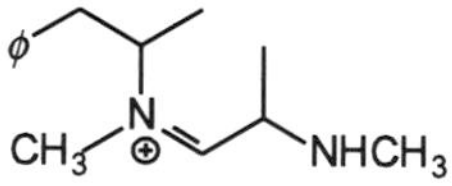

I

e. *m/z* 205 (loss of φCH$_2$·):

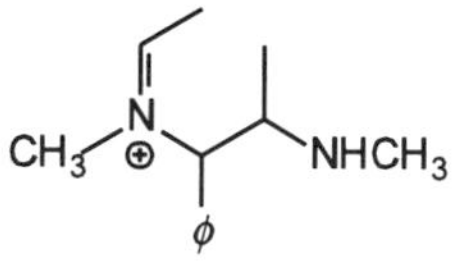

J

f. *m/z* 58:

K

Structures B and E (both $m/z$ 295) should be the least stable of this group since the isolated terminal double bonds have no stabilizing features. Somewhat more stable, by virtue of having isolated interior double bonds, are structures A and D ($m/z$ 295), F and G ($m/z$ 281), I ($m/z$ 219), J ($m/z$ 205), and K ($m/z$ 58).

The most stable ion structures are C ($m/z$ 295) and H ($m/z$ 238), both of which contain interior double bonds conjugated with an aromatic ring. Because the radical product formed with H is more stable than hydrogen radical, $m/z$ 238 should be the base peak in the spectrum. However, the radical products formed along with ions J and K are also resonance-stabilized. If the difference in ion stability is less than the difference between the stabilities of the corresponding radical products, it is possible that $m/z$ 205 or 58 could become the base peak. In any case, all three ions should be prominent in the spectrum. Because of the stability of C ($m/z$ 295), this ion may be fairly intense as well.

5.5. An important piece of information is included with these two unknowns—namely, that they were recovered from a clandestine drug lab that was manufacturing methamphetamine. Even without knowledge of the exact synthetic method used, we can assume that the structures of these unknowns *might* be related to that of methamphetamine.

a. The isotopic abundance data for the ions at $m/z$ 134 and 43 in Figure 5.26*a* are helpful: $m/z$ 134 appears to contain nine carbons, while $m/z$ 43 seems to contain just two. This means that $m/z$ 43 is probably the acetyl ion ($CH_3C\equiv O^+$), not $CH_3CH_2CH_2^+$. The ion at $m/z$ 91, on the other hand, seems consistent with benzyl ($\phi CH_2^+$) despite the lack of reliable isotopic abundance information, due to the $91 \rightarrow 65 \rightarrow 39$ low mass ion series. If this is true, and if the assumption that the molecular ion contains nine carbons is correct, then this molecule is made up of two fragments—benzyl and acetyl ($91 + 43 = 134$). There is only one reasonable arrangement of these fragments that fits the data: methylbenzylketone, also known as phenyl-2-propanone, or P-2-P, which has obvious structural similarites to methamphetamine. [*Answer:* Phenyl-2-propanone]

(9.18)

m/z 119          m/z 134          m/z 91

CH₃—C≡O ⊕

m/z 43

b. The second unknown has a relatively high molecular weight, which under some circumstances might make its solution difficult. Two factors help here:

First, the spectrum is remarkably simple (the first loss is 91 daltons to give the ion at $m/z$ 119), and second, the structure of this compound may be related to that of P-2-P (Unknown 5.5$a$).

The $91 \rightarrow 65 \rightarrow 39$ low mass ion series strongly suggests the presence of benzyl; in fact, there is no indication of any group other than benzyl in this molecule! It is not unreasonable to assume that the molecular ion loses benzyl to form $m/z$ 119. Further, if the $m/z$ 119 ion does not contain a benzyl group, there are no other clues as to what this ion might contain since the entire spectrum up to $m/z$ 100 is that of a benzyl group and its fragments. If we postulate the presence of two benzyl groups in this compound, we are left with only $210 - (2 \times 91) = 28$ daltons to account for. Although this group could be due either to CO or $CH_2CH_2$, the isotopic abundance information in the molecular ion cluster is consistent with 15 carbons since $[(M + 1)^+]/[M^+] = 16.6\%$—two benzyl groups plus an additional carbon. A carbonyl group is also consistent with the presence of P-2-P in the sample and gives a strong directing group for fragmentation as well. The ion at $m/z$ 118 in this spectrum, as well as that at $m/z$ 92 in the spectrum of P-2-P, are both due to the $\gamma$-hydrogen rearrangement, which is discussed in Section 6.2. [*Answer:* Dibenzylketone]

(9.19)

m/z 210                    m/z 119                    m/z 91

**5.6.** As with several previous problems, it is instructive first to list all possible structures and try to predict beforehand how each will fragment. In this case we need to consider not only primary $\alpha$-cleavages, but also secondary fragmentations of the initial $\alpha$-cleavage ion as well as other fragmentations that are unique to aliphatic oxygen compounds. Seven structures are possible, as shown in Table 9.2.

Two of these spectra should be readily identifiable. The spectrum in Figure 5.27$a$ is the only one showing loss of water ($m/z$ 56), so it must belong to $n$-butanol. Also, that in Figure 5.27$g$ shows loss of a methyl group with no corresponding loss of hydrogen radical, consistent with behavior expected for $t$-butyl alcohol. Of the remaining spectra, Figures 5.27$b$ and $e$ exhibit large ions at $m/z$ 45. Since Figure 5.27$e$ also shows a significant loss of methyl, it must correspond to $sec$-butyl alcohol, and Figure 5.27$b$ (Problem 4.6) must be that of methyl-$n$-propylether.

Beyond this, it is less clear into what "base peak category" some of the remaining spectra fall. Figure 5.27$c$ seems clearly to be in the $m/z$ 59 category and Figure 5.27$f$ in the $m/z$ 31 column, but the base peak in Figure 5.27$d$ is

**Table 9.2. Predicted Losses of $C_4H_{10}O$ Isomers**

| | Structure | Primary α-Cleavage Losses[a] | Secondary Losses |
|---|---|---|---|
| n-Butyl | | 2 H˙ (m/z 73); CH₃CH₂CH₂˙ (m/z **31**) | (Loss of $H_2O$) |
| sec-Butyl | | H˙ (m/z 73); CH₃˙ (m/z 59); CH₃CH₂˙ (m/z **45**) | None |
| Isobutyl | | 2 H˙ (m/z 73); CH₃ĊHCH₃ (m/z **31**) | None |
| t-Butyl | | 3 CH₃˙ (m/z **59**) | None |
| Methylpropyl | | 5 H˙ (m/z 73); CH₃CH₂˙ (m/z **45**) | 73→31 |
| Methylisopropyl | | 4 H˙ (m/z 73); 2 CH₃˙ (m/z **59**) | 73→31 |
| Diethyl | | 4 H˙ (m/z 73); 2 CH₃˙ (m/z **59**) | 73→45 59→31 |

[a]Most important a-cleavage loss shown in **boldface**.

m/z 43, which is not due to charge retention α-cleavage if the compound is indeed an aliphatic alcohol or ether. Since the ions at m/z 45 and 59 are so weak in this spectrum, the modest peak at m/z 31 must take precedence by default, and isobutyl alcohol becomes the answer for this spectrum. This has some merit since loss of an isopropyl group by α-cleavage is predicted, although in this case it occurs primarily by charge migration, rather than charge retention. Because of the high electronegativity of oxygen, the incipient secondary isopropyl carbocation apparently has a more stablizing effect than the unsubstituted $CH_2=OH^+$ ion. (In contrast, the spectrum of isobutylamine has a base peak at m/z 30 that is over 10 times larger than any other ion in the spectrum!)

Of the two remaining spectra, Figure 5.27c seems most clearly attached to methylisopropylether, since m/z 45 is entirely absent from the spectrum. This leaves the assignment of diethylether to Figure 5.27f, which at first seems surprising since m/z 31 is the base peak. Upon reflection, however, this is consistent with the fact that the secondary fragmentations from initial α-cleavage ions are more prominent in ethers and alcohols than in amines.

(9.20)

m/z 45          m/z 73          m/z 74          m/z 59          m/z 31

It should be clear that without having the spectra of all of these isomers to compare and contrast, unique structural assignments would be much more difficult. [*Answer:* (a) *n*-Butanol, (b) methylpropylether, (c) methylisopropylether, (d) isobutyl alcohol, (e) *sec*-butyl alcohol, (f) diethylether, and (g) *t*-butyl alcohol]

**5.7.** a.

b.

c.

d.

e.

f.

g.

h.

i.

j.

**5.8.** At first glance, there seems to be conflicting data in this spectrum. For one thing, the isotopic abundance data for the $m/z$ 122 and 106 ions both indicate the presence of seven carbon atoms. From the low mass ion series at $m/z$ 39, 51, and 78, we expect the compound to be aromatic. If the aromatic ring is a benzene ring, this leaves only one extra carbon and a lot of mass unaccounted for. On the other hand, the first loss from the molecular ion is fairly unusual (16 daltons), one observed mainly in primary amides and a few other selected compounds (Table 4.1). In either case, the presence of nitrogen is indicated by this loss, yet the molecular weight is even! It is important to remember, however, that the odd-nitrogen rule states that a compound having an even molecular weight must have an *even* number of nitrogens. In essentially every case so far the even number has been 0, but in this case that assumption does not work, and the next higher number (2) seems more reasonable.

This compound thus appears to be an aromatic primary amide, an assumption that is corroborated by the presence of a relatively weak $m/z$ 44 ion that is not intense enough to be from an aliphatic amine, but which is expected in the spectra of primary amides ($^+CONH_2$; Figure 5.19). In fact, comparing this spectrum with that of benzamide (Figure 5.19$b$), we notice that the two mass spectra are nearly identical except that the intense ions at $m/z$ 77, 105, and 121 in the benzamide spectrum occur at masses one dalton higher in the spectrum of this compound. Further, the loss from $m/z$ 78 to 51 is not the usual 26 daltons (HCCH), but rather 27 (HCN) indicating the presence of a nitrogen in the aromatic ring (which also accounts for the "missing" nitrogen atom). A pyridine ring fulfills the necessary requirements and also accounts for all of the remaining mass. The intensities of the $m/z$ 123 and 124 ions thus can be explained by the contributions of six carbons and two nitrogens. [*Answer:* 3-Pyridinecarbox-amide (nicotinamide)]

(9.21)

m/z 106  m/z 122  m/z 44

(9.22)

m/z 122  m/z 78  m/z 51

**5.9.** The odd molecular weight and base peak at *m/z* 72 are giveaways that this is the spectrum of an aliphatic amine. In fact, except for the molecular ion, nearly all of the major ions in the spectrum are at masses expected for aliphatic amine ions (*m/z* 30, 44, 58, 72, and 86), and the isotopic abundance data for *m/z* 72 is consistent with this fact.

Eighteen structures are possible for the empirical formula $C_5H_{13}N$. Most of these can be eliminated quickly by considering what the base peak at *m/z* 72 really tells us—namely, that methyl is the *largest* group that can be lost by initial α-cleavage (secondary rearrangements give rise to less intense ions in aliphatic amines). Thus *m/z* 44 cannot arise from loss of a propyl group; if this were true, *m/z* 44 would be larger than *m/z* 72, since propyl radical is more stable than methyl (Section 5.3). This leaves only four structures to consider: methyl *t*-butylamine, ethylisopropylamine, dimethylisopropylamine, and methyldiethylamine. These structures can be distinguished by looking at the pattern of secondary rearrangements from α-cleavage ions that each produces (see Table 9.3).

Methyl *t*-butylamine and dimethylisopropylamine can be eliminated immediately on this basis, since each is predicted to produce only one secondary rearrangement ion, instead of the three observed in this spectrum. Choosing between ethylisopropylamine and methyldiethylamine is more subtle. However, the base peak (*m/z* 72) for ethylisopropylamine should produce two intense secondary rearrangement ions—at *m/z* 44 and 30—depending on which methyl group is lost. The same is not true for methyldiethylamine. Thus the relative intensities of the secondary rearrangement ions for this compound indicate that, although *m/z* 44 arises from *m/z* 72, *m/z* 30 probably does not, and that this pat-

Table 9.3. Possible Solutions for Problem 5.9

| Structure | α-Cleavage Ion | Secondary Rearrangements |
|---|---|---|
| (amine structure) | m/z 86  \n m/z 72 | 86→30  \n None |
| (amine structure) | m/z 86  \n m/z 72 | 86→44   86→58  \n 72→44   72→30 |
| (amine structure) | m/z 86  \n m/z 72 | 86→44  \n None |
| (amine structure) | m/z 86  \n m/z 72 | 86→58→30  \n 72→44 |

tern better fits that expected of methyldiethylamine than of ethylisopropylamine. [*Answer:* Methyldiethylamine]

## CHAPTER 6

**6.1.** This spectrum has several features of saturated aliphatic hydrocarbon spectra - intense ions at $m/z$ 43 and 57 and an apparent molecular weight of 100. A more careful examination of the data, however, indicates something different. First, the isotopic abundance information for the $m/z$ 43, 57, and 72 ions is consistent with the presence of two, four, and four carbons, respectively, meaning at least that $m/z$ 43 must be the acetyl ion, not $C_3H_7^+$. On the other hand, $m/z$ 57 is $C_4H_9^+$. Since acetyl is only observed as the base peak in carbonyl compounds in which it is a terminal functional group (i.e., it is formed by α-cleavage and not by rearrangement from interior groups in molecules), it is not unreasonable to assume that acetyl and $C_4H_9^+$ account for the entire molecule: $CH_3COC_4H_9$.

Four isomeric structures fit this formula, and the rest of the spectrum must be used to help distinguish between them. As we have seen, aliphatic ketones undergo the γ-hydrogen (McLafferty) rearrangement under appropriate conditions. We can narrow down the possibilities by looking at these structures and predicting the products of McLafferty rearrangement for each of them (Table 9.4) Since the observed rearrangement ion occurs at $m/z$ 72 (the only odd-electron ion in the spectrum other than the molecular ion), only 3-methyl-2-pentanone is consistent with the observed spectrum. [*Answer:* 3-Methyl-2-pentanone (*sec*-butyl methyl ketone)]

**Table 9.4. Possible Solutions to Problem 6.1**

| | Predicted McLafferty Ion | | Predicted McLafferty Ion |
|---|---|---|---|
| | *m/z* 58 | | *m/z* 72 |
| | *m/z* 58 | | No γ-hydrogen |

**6.2.** Although $C_4H_9$ also has a mass of 57 daltons and could conceivably produce the base peak, cyclohexanol itself has a base peak at *m/z* 57 due to $C_3H_5O^+$:

$$(9.23)$$

**6.3.** Both the phenyl group and the nitrogen in the piperidine ring act as directing groups for α-cleavage within the cyclohexane ring. This is the structural requirement necessary to initiate the cyclohexanone-type rearrangement:

$$(9.24)$$

**6.4.** From a purely naïve viewpoint, these two compounds can be distinguished by identifying the products each produces in the retro Diels–Alder fragmentation:

4-Hydroxycyclohexene:                                                                (9.25)

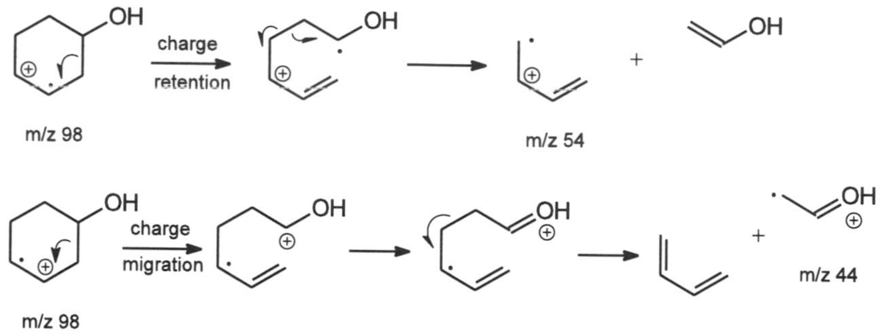

charge
retention

m/z 98

m/z 98   m/z 54   +

charge
migration

m/z 98   m/z 44   +

3-Hydroxycyclohexene:                                                                (9.26)

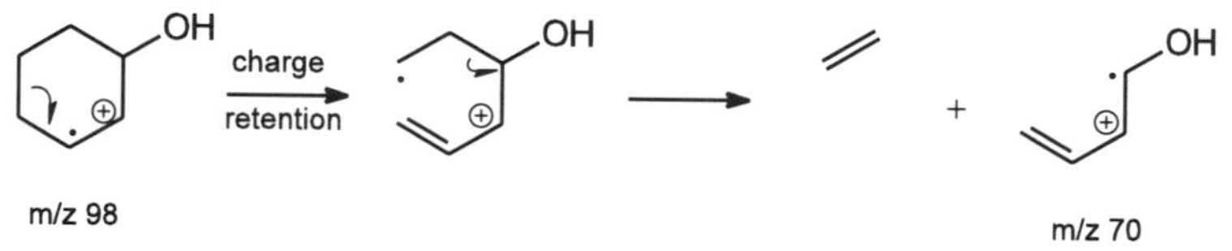

charge
retention

m/z 98   m/z 70   +

Even though this approach produces the correct answer, the actual situation is much more complex since initial ionization is expected to occur on oxygen, rather than at the double bond. In fact, this assumption helps explain some other fragmentations of these two compounds that otherwise seem enigmatic:

3-Hydroxycyclohexene:                                                                (9.27)

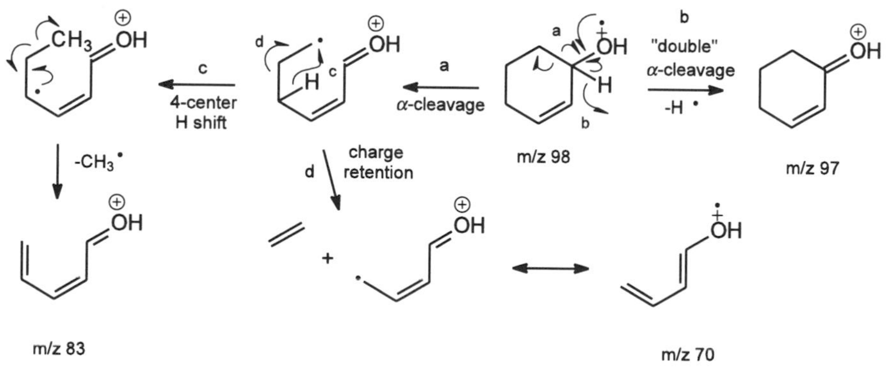

$CH_3$   OH

c
4-center
H shift

a
α-cleavage

a   b
"double"
α-cleavage

-H ·

charge
retention

d

-$CH_3$ ·

m/z 98

m/z 97

m/z 83   +   m/z 70

The 3-hydroxy isomer readily loses the hydrogen that is next to both oxygen and a double bond ("double α-cleavage"), as well as a methyl group via a mechanism that resembles the methyl loss in cyclohexanone (Figure 6.11). The

(9.28)

4-hydroxycyclohexene:

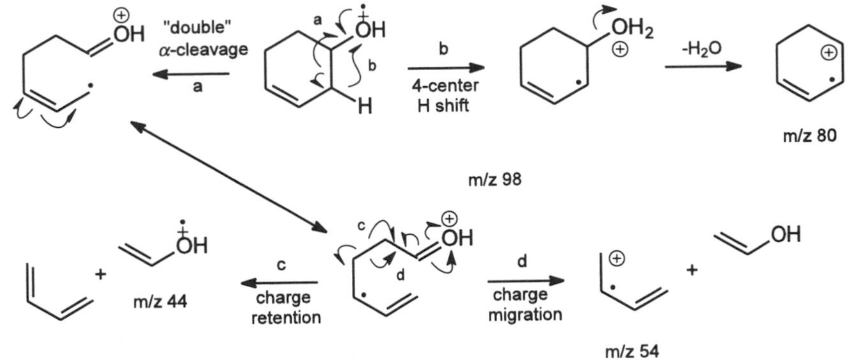
"double"
α-cleavage
a
b
4-center
H shift
-H₂O
m/z 80
m/z 98
c
charge
retention
m/z 44
d
charge
migration
m/z 54

4-hydroxy isomer, on the other hand, loses water after an initial (and relatively facile) migration of an allylic hydrogen. Overall, the behavior of both of these compounds more closely resembles that of cyclohexanone-type compounds than of cyclohexene derivatives. [*Answer:* (a) 4-Hydroxycyclohexene and (b) 3-hydroxycyclohexene]

**6.5.** Both of these compounds can undergo the retro Diels–Alder fragmentation [compare cannabidiol; Figure 6.18 and Eq. (6.21)], but only the $\Delta^8$-isomer leads to the actual loss of a fragment. In the $\Delta^9$-isomer everything is still attached following the rearrangement.

$\Delta^8$-THC: (9.29)

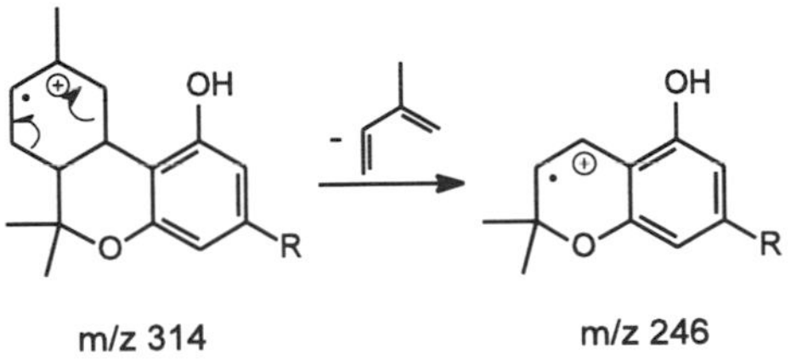
OH
R
m/z 314
m/z 246

$\Delta^9$-THC: (9.30)

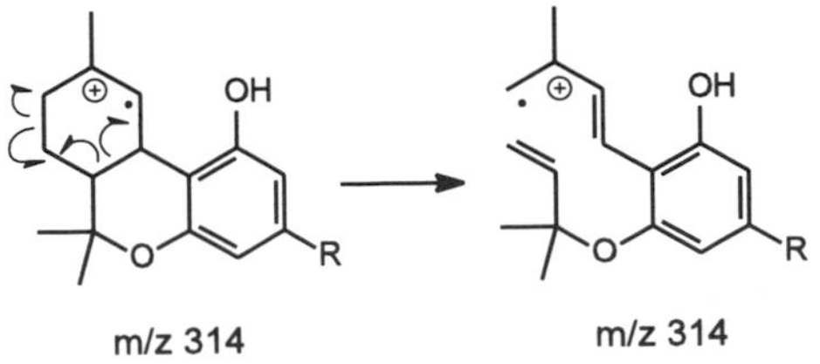
OH
R
m/z 314
m/z 314

## CHAPTER 7

**7.1.** The products of α-cleavage are formed directly by breaking the bond between the carbonyl carbon and the carbon next to nitrogen. It is difficult to envision re-arrangement processes that would lead to other products in this molecule since loss of water would involve rearrangement of two hydrogens and form a vinylic carbocation. The fact that even minor peaks in both spectra are explainable on the basis of their structures means that mass spectrometry, especially when cou-pled with a complementary chromatographic method, is an acceptable method for identifying these two compounds.

(9.31)

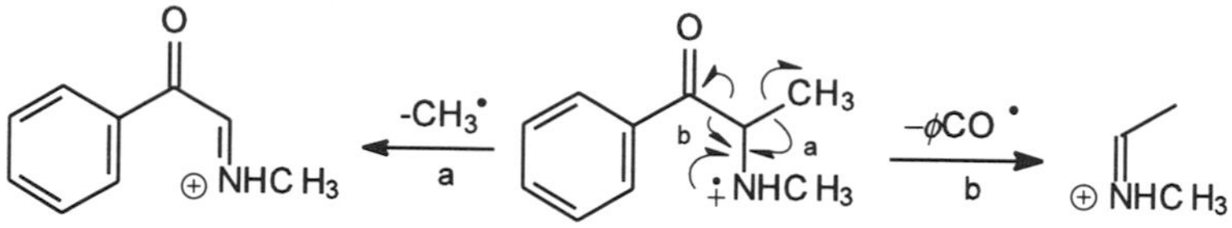

m/z 148                m/z 163                m/z 58

(9.32)

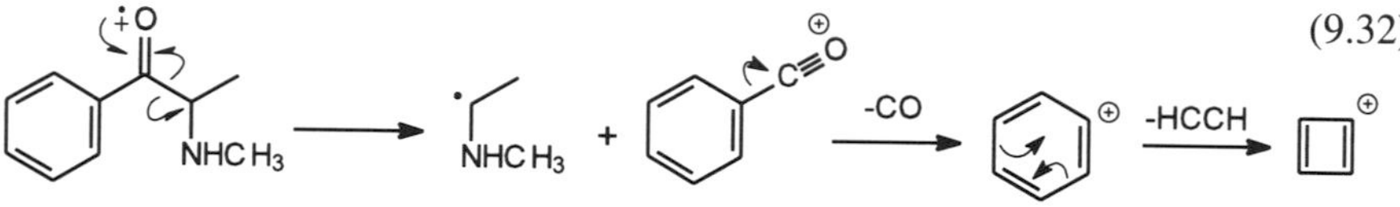

m/z 105                m/z 77                m/z 51

**7.2.** The presence of an *ortho* methyl group in the 2,5-dimethyl isomer makes possi-ble a rearrangement via a 6-center transition state:

(9.33)

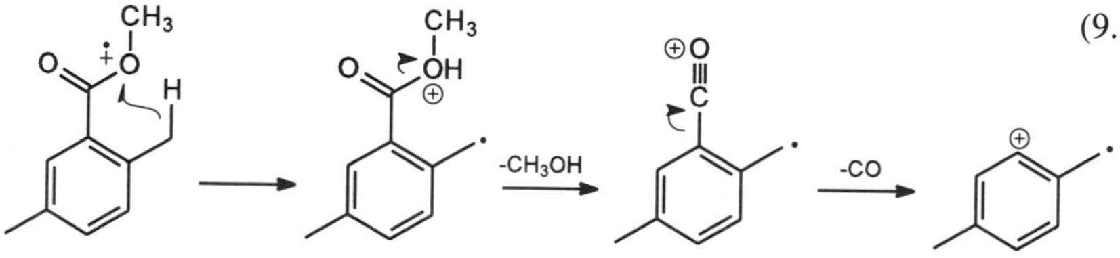

m/z 164                m/z 132                m/z 104

(Accounting for the loss of CO from m/z 132 is tricky since the carbon in CO has a lone pair of electrons attached to it!)

**7.3.**                                                                            (9.34)

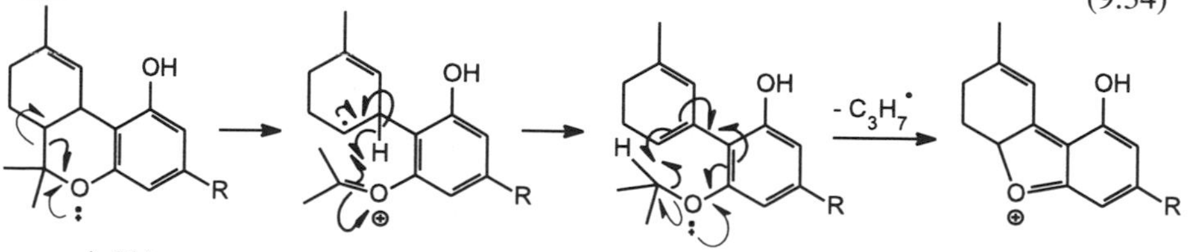

m/z 314                                              m/z 271

In order to lose the *gem*-dimethyl group, its attached carbon, and an additional hydrogen, three things must happen. First, one of the bonds to the attached carbon must break. The resultant radical or ion then must provide a site toward which transfer of an appropriate hydrogen can take place. Finally, cleavage of a second bond to the attached carbon must free the $C_3H_7$ fragment from the rearranged molecular ion and, because this ion is reasonably intense, must generate a structure stabilized by extended conjugation (compare the structures of the other high mass ions in this spectrum; Figure 6.7). [Equation (9.34) is adapted with permission from Smith, 1997. Copyright ASTM.]

After initial ionization at the ring oxygen, breaking the bond between the attached carbon and the cyclohexene ring (α-cleavage) places a radical site in the cyclohexene ring and the positive charge on the oxygen. Rearranging hydrogen to either the cyclohexane ring or to the oxygen gains nothing toward the eventual loss of the $C_3H_7$ fragment. However, a hydrogen six atoms away from the *gem*-dimethyl carbon should migrate easily to the desired site. Transfer of this hydrogen to the *gem*-dimethyl carbon is accompanied by (a) formation of a second double bond in the cyclohexene ring (this one conjugated with the aromatic ring) and (b) reformation of the radical ion on the oxygen. This ion should be similar in stability to the unrearranged molecular ion, thus keeping energy demands to a minimum. Although the final step in this rationalization involves cleavage of a carbon–heteroatom bond, this is balanced by formation of a new carbon–oxygen bond and of an ion possessing substantial conjugation [compare Eq. (5.12)].

**7.4.** The methyl group on the pyridine ring is a benzylic carbon; its loss would not be expected (Section 5.2). Loss of the ether methyl group by long-distance α-cleavage, on the other hand, has precedent [e.g., Eq. (5.12)], as does the subsequent loss of carbon monoxide [Eq. (4.4) and Figure 6.11]. Loss of 28 daltons as ethylene, rather than as CO, seems unlikely in such a highly aromatic system (see Section 4.2.2).

$$(9.35)$$

m/z 212                    m/z 197                    m/z 169

**7.5.** The fragment ions at $m/z$ 209, 194, and 180 all appear to contain the chlorine, and undoubtedly the aromatic ring as well. The aliphatic nitrogen atom should, based on previous experience, direct fragmentation that will retain the positive charge on the nitrogen (Section 5.3). Not surprisingly, this means that the part of the molecule most likely to fragment is the cyclohexanone ring. Consider the fragility of the bond between the carbonyl carbon and the carbon containing both

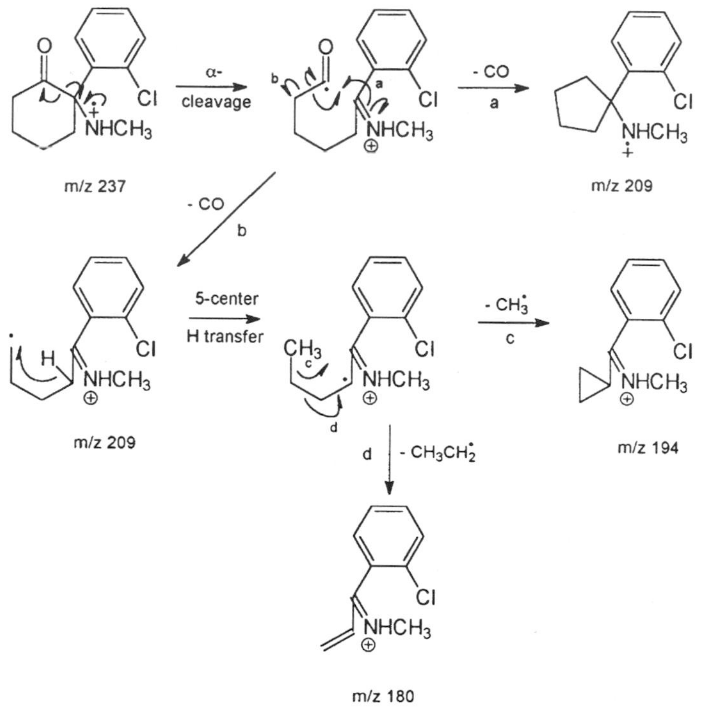

**Figure 9.2.** Proposed fragmentations of ketamine (Problem 7.5).

the amine and aromatic ring; this is a bond attached to a carbonyl carbon, a carbon next to nitrogen and a benzylic carbon ("triple" α-cleavage)! Little wonder that the molecular ion at $m/z$ 237 is so small.

Combining all of this information leads to initial ionization at nitrogen, followed by α-cleavage of the fragile bond and stabilization of the positive charge on nitrogen (Figure 9.2). The situation is now similar to that seen after initial cleavage in both cyclohexanone (Figure 6.11) and phencyclidine (Problem 6.3; see also Section 8.4.1). Loss of carbon monoxide to produce $m/z$ 209 can occur with relative ease, as in cyclohexanone (Figure 6.11), or with ring closure to regenerate a radical ion similar in stability to that of the molecular ion. Generation of a primary radical site, on the other hand, should lead to further fragmentation. One possibility is a cyclohexanone-type rearrangement involving a 5-center, rather than 6-center, transfer of hydrogen to move the radical site nearer to the nitrogen. Subsequent loss of methyl and ethyl radicals lead to the ions at $m/z$ 194 and 180, respectively. As in cyclohexanone (Figure 6.11), loss of the larger radical is greatly preferred because both the radical and the resultant ion products are considerably more stable.

## CHAPTER 8

**8.1.** This spectrum is very similar to that of 3,4-methylenedioxyethylamphetamine (MDEA; Figure 8.2c), *but it is not identical!* Most notably, the peak at *m/z* 44 in the MDEA spectrum is absent here. However, the presence of the *m/z* 135 ion, the base peak at *m/z* 72 and the apparent molecular ion at *m/z* 207 strongly suggests that this compound is an isomer of MDEA that differs only in the arrangement of the carbons near the nitrogen:

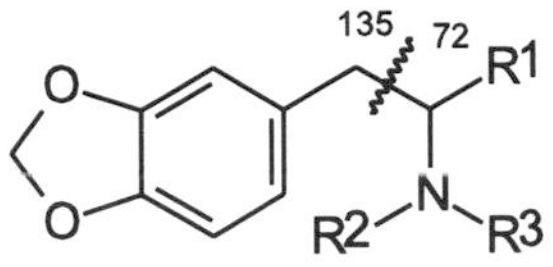

Eight additional structures meet these requirements. The pattern of losses observed in the unknown spectrum should help us decide among these possibilities (Table 9.5).

   This unknown loses hydrogen to produce *m/z* 206 and methyl to give *m/z* 192 (remember that some α-cleavage losses in these compounds may produce ions of extremely low abundance based on the relative stability of the radicals formed). It does not appear to lose either an ethyl or a propyl group (the ions at *m/z* 178 or 164 are insignificant in size), eliminating structures C, D, E, and F. Although the spectrum lacks the *m/z* 44 ion present in the spectrum of MDE, thereby ruling out structure H, we cannot be certain if another possible secondary rearrangement ion (namely, *m/z* 30) is present or not, since the spectrum has not been recorded below *m/z* 40. Although structures B and G cannot be ruled out by the available data, in both cases we might expect *m/z* 192 to be substantially larger than *m/z* 206 (compare methamphetamine and phentermine; Figure 5.10). [*Answer: N,N*-dimethyl-3,4-methylenedioxyamphetamine (structure A)]

**8.2.** Phenylacetylmethylecgonine and toluylmethylecgonine should have extremely similar spectra—so similar, in fact, that they may be distinguishable only by comparing the spectra of known standards. Since the only structural difference is in the aryl groups, only the abundances of ions directly involving fragmentation of these groups should be affected to any extent. For example, the ions at *m/z* 136 (the "benzoic acid" ion), 119 (the "benzoyl" ion), and 91 ("benzyl" ion) will have different structures when these two compounds fragment, so that they, and their fragment ions at *m/z* 65 and 39 (the benzyl low mass ion series), will probably differ somewhat in relative intensity between the two spectra. But even these differences may be minimal. Unfortunately, no mass spectrum was available for comparison.

**8.3.** The intense ions at *m/z* 82 and 196 are sufficient to identify this spectrum as a derivative of cocaethylene (compare Figure 8.6a). The loss of 45 daltons from

**Table 9.5. Isomeric Structures for Problem 8.1**

| | | α-Cleavages | Secondary rearrangements |
|---|---|---|---|
| A. | | H˙ <br> CH$_3^+$ | None |
| B. | <br> NHCH$_3$ | H˙ <br> 2 CH$_3^+$ | None |
| C. | <br> NH$_2$ | CH$_3^+$ <br> CH$_3$CH$_2^+$ | None |
| D. | <br> NH$_2$ | H˙ <br> CH$_3$CH$_2$CH$_2^+$ | None |
| E. | <br> NH$_2$ | H˙ <br> CH$_3^+$CHCH$_3$ | None |
| F. | <br> HN | H˙ <br> CH$_3$CH$_2^+$ | m/z 30 |
| G. | <br> HN | H˙ <br> 2 CH$_3^+$ | m/z 30 |
| H. | <br> N | H˙ <br> CH$_3^+$ | m/z 44 |

the molecular ion ($m/z$ 363 → $m/z$ 318) also is consistent with the presence of the ethyl ester group, while presence of the $m/z$ 82 ion confirms that the ring system near, and including, the $N$-methyl group is intact (Figure 9.3).

The nature of the aryl group remains to be determined. The loss of 151 daltons from the molecular ion to give the ion at $m/z$ 212 is duplicated by the presence of an ion at $m/z$ 151, corresponding, respectively, to loss of the substituted

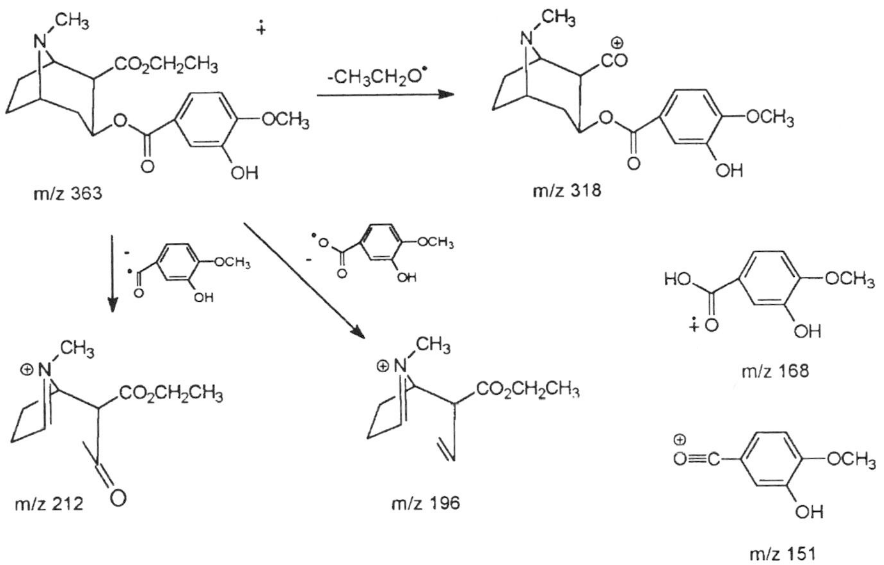

**Figure 9.3.** Proposed fragmentations of hydroxymethoxycocaethylene (Problem 8.3).

benzoyl ion and to the substituted benzoyl ion itself. Similarly, the loss of the substituted benzoate radical (167 daltons to give $m/z$ 196) is mirrored in the presence of the benzoic acid ion at $m/z$ 168. Comparison with Figure 8.6c and Table 8.1 shows that $m/z$ 151 and 168 are prominent in the spectra of hydroxymethoxycocaines and correspond to the substituted benzoyl ion and benzoic acid radical ion, respectively (Figure 9.3). [*Answer:* Hydroxymethoxycocaethylene]

**8.4.** The list of ions expected for *N,N*-diethyl-1-phenylcyclohexylamine is given below. The ion structure types are those in Figures 8.15 to 8.17. The losses of methyl and ethyl radicals are a by-product of the cyclohexanone-type rearrangement (see Figure 6.11) and are seen in the spectrum of phencyclidine (Figure 8.14).

| Ion Structure | Fragmentation | Predicted Ion |
|---|---|---|
| — | Isotopic abundance ion | 232 (5%) |
| XVIII | Molecular ion | 231 (25%) |
| XIXb | Loss of H$^\bullet$ from the phenyl ring | 230 (30%) |
| — | $-CH_3^\bullet$ from rearrangement intermediate | 216 (1%) |
| — | $-CH_2CH_3^\bullet$ from rearrangement intermediate | 202 (2%) |
| — | Isotopic abundance ion | 189 (15%) |
| XX | Cyclohexanone-type rearrangement | 188 (100%) |

| Ion Structure | Fragmentation | Predicted Ion |
|---|---|---|
| — | Isotopic abundance ion | 175 (3%) |
| XXI | Loss of H• + $C_4H_8$ | 174 (20%) |
| XXIII | Formation of phenylcyclohexene | 158 (5%) |
| — | Isotopic abundance ion | 155 (2%) |
| XXII | Loss of phenyl radical | 154 (15%) |
| XXIV | Retro Diels–Alder from XXIII | 130 (5%) |
| XXV | — | 117 (10%) |
| — | Loss of H• from XXV | 115 (9%) |
| XXVI | — | 104 (7%) |
| — | Loss of H• from XXVI | 103 (6%) |
| — | Isotopic abundance ion | 92 (3%) |
| XXVII | — | 91 (35%) |
| — | Phenyl ion | 77 (8%) |
| XXVIII | Loss of phenylcyclohexene from XIXb | 72 (16%) |

A comparison of the predicted and actual spectra of this compound can be seen in Figure 9.4.

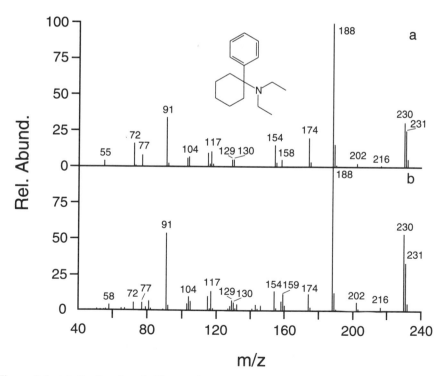

**Figure 9.4.** (*a*) Predicted and (*b*) actual spectra of *N,N*-diethyl-1-phenylcyclohexylamine (Problem 8.4).

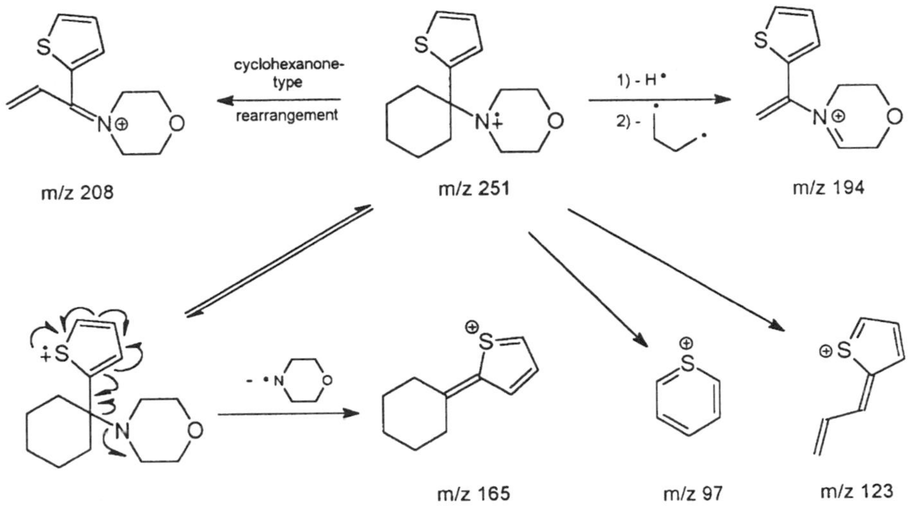

**Figure 9.5.** Proposed fragmentations of 1-(2-thienyl)cyclohexylmorpholine (Problem 8.5).

**8.5.** The intense ions at $m/z$ 97 and 165 and smaller one at $m/z$ 123 are all found in the spectrum of the thiophene analog of phencyclidine (Figure 8.18c). In fact, the ion series at $m/z$ 81, 109 and 110, 135 and 136, and 149 to 150 are also found in both spectra. In addition, the major loss of 43 daltons from the molecular ion, coupled with a smaller loss of (56 + 1) daltons to give $m/z$ 194, is typical of phencyclidine derivatives containing the cyclohexane ring. Thus it seems likely that this compound differs from 1-(2-thienyl)cyclohexylpiperidine by replacement of the piperidine ring with some other group. Although the nature of the unknown group may not be obvious, it is helpful to notice that the molecular weight of this compound, as well as the ions at $m/z$ 208 and 194, all differ from those in the spectrum of 1-(2-thienyl)cyclohexylpiperidine by 2 daltons. A similar difference is seen between the spectra of phencyclidine and phenylcyclohexylmorpholine (Figures 8.14 and 8.18b), so that a morpholine ring is a good candidate for the missing group (Figure 9.5). [*Answer:* 1-(2-Thienyl)cyclohexylmorpholine]

**8.6.** Instead of losing phenyl radical by α-cleavage, this compound loses a benzyl group, thereby producing both a stable ion and a very stable radical. The energy of activation for this one-step process must be substantially lower than that for the multistep cyclohexanone-type rearrangement.

(9.36)

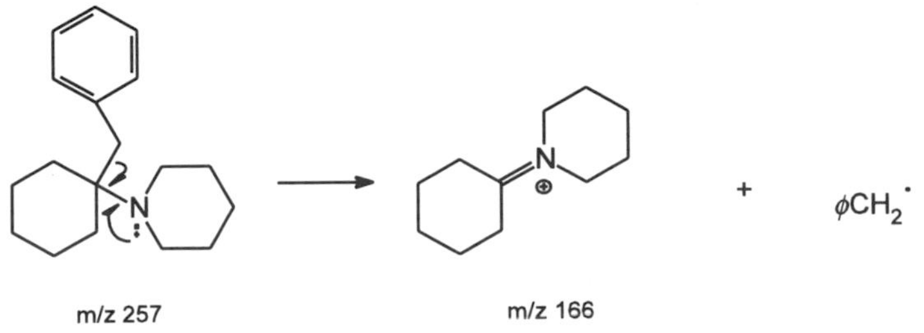

$m/z$ 257                    $m/z$ 166

**8.7.** The low mass ion series provides clues about the general structure of this unknown—the ions at $m/z$ 59, 70, 81, 115, 124, and 162 are typical of the morphine alkaloids in this section. The intense $m/z$ 43 ion also indicates that this is an acetylated morphine derivative (compare Figures 8.20 and 8.22). A little arithmetic suggests a monoacetylated morphine structure. Although many acetylated morphines are possible, the two $O$-acetyl derivatives would be the easiest to synthesize and thus the most logical structures with which to start. The question then becomes which of the two isomers it is.

The two acetoxy groups in heroin each fragment in a characteristic manner—the aromatic acetoxy by cyclic loss of ketene and the allylic one by loss of the entire acetoxy group. In this spectrum loss of 42 daltons from the molecular ion to give the base peak at $m/z$ 285 indicates loss of ketene, meaning that the acetyl group must be attached to the oxygen on the aromatic ring. This derivative is, in fact, the less common monoacetylmorphine. The 6-acetyl isomer, which fragments to give a base peak at $m/z$ 268 (327 − 59), is the usual metabolite and the normal partial hydrolysis product of illicit heroin samples. [*Answer:* 3-Acetylmorphine]

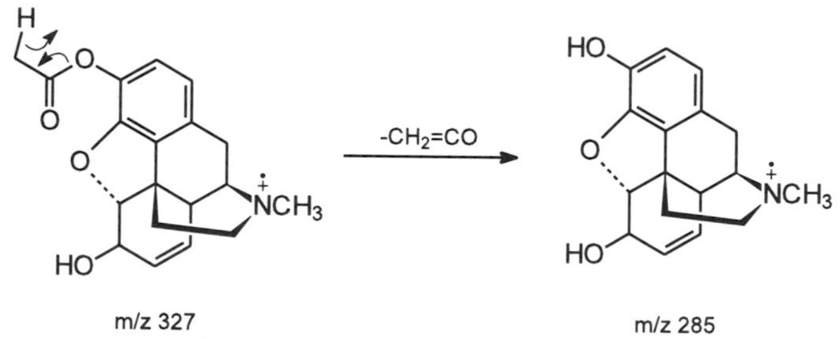

(9.37)

m/z 327          m/z 285

**8.8.** The high mass ions in the spectrum of compound B can be readily understood on the basis of previously discussed fragmentations:

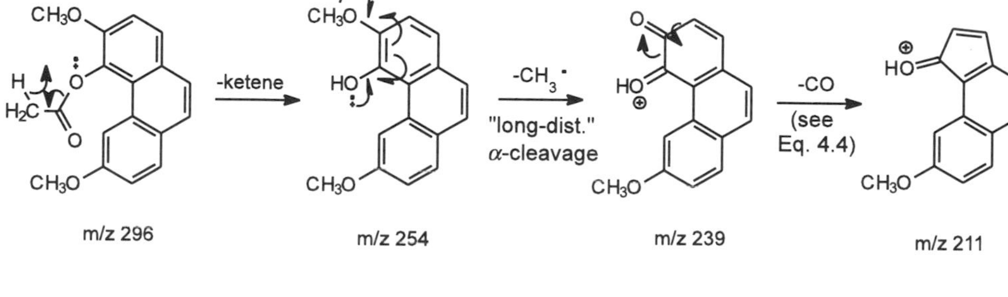

(9.38)

m/z 296          m/z 254          m/z 239          m/z 211

Unknowns C and D show similar high mass losses: sequential losses of 42 daltons (two for unknown C and three for unknown D), then loss of a methyl group by C and hydrogen radical by D (in both instances giving $m/z$ 225) followed by loss of 28 daltons to give $m/z$ 197. By comparison with the behavior of compound B, it seems likely that C contains two similar acetyl groups and a methyl group that can be lost, while D has three acetyl groups but no fragmentable methyl group. Structures that fulfill these requirements are similar to that of acetylthebaol (B) but are derived from codeine (with two OH groups capable of derivatization by acetyl plus a methyl group on the remaining oxygen) and morphine (three derivatizable —OH groups and no methyl), respectively:

Peak C                                                  Peak D

The mass spectrum from peak A, on the other hand, is more difficult to interpret. In this case, no loss of ketene is observed, thus there are apparently no acetyl groups in the molecule. Loss of a methyl group produces an intense ion at $m/z$ 237, but the peak at $m/z$ 220 is of uncertain origin, since it could come from loss of 32 daltons (as MeOH?) from the molecular ion or of 17 (as OH?) from $m/z$ 237. The only clues about the molecular ion come from the fact that 252 is 2 daltons lower than the mass of the base peak in the spectrum of compound B (i.e., the deacetylated acetylthebaol radical ion). Thus compound A could be related to unacetylated thebaol in having an extra double bond (although this seems highly unlikely given the high degree of unsaturation already present in the molecule) or an extra ring. The latter could be achieved by forming a five-membered oxygenated ring, using the unmethylated oxygen of thebaol, like that observed in morphine and codeine. This could explain the favorable loss of methyl, although formation of the peak at $m/z$ 220 remains unclear:

(9.39)

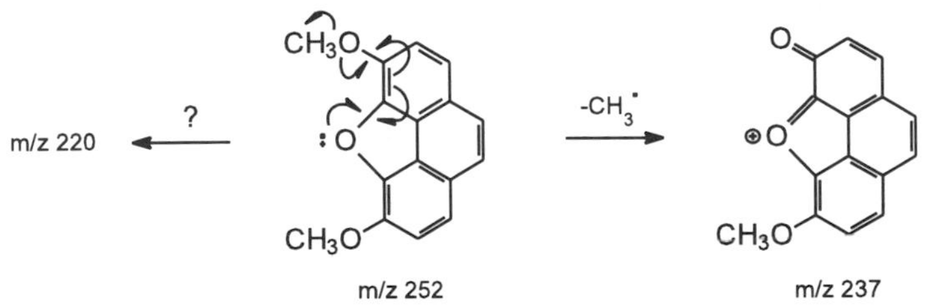

$m/z$ 220 ← ? —         —$CH_3^{\bullet}$ →

m/z 252                        m/z 237

Although the above structures reasonably account for most of the data, so little sample was available that it could not be analyzed by other spectroscopic methods. Thus these answers are simply "educated guesses." The origin of these compounds remains an even greater mystery. This was a completely unique forensic heroin sample. It is possible that opium was mixed with aspirin as a home pain remedy, a practice used by some southeast Asian cultures (Smith and

Nelsen, 1991). Intense heating of the opium/aspirin mixture would undoubtedly decompose all of the thebaine (which is thermally labile in any case) and possibly some of the morphine and codeine. It would certainly decompose the aspirin, which is an extremely good acetylating agent that leaves no trace of its presence after the reaction is over (the resulting salicylic acid sublimes out of the mixture!).

**8.9.** The five unknown spectra can be arranged into two groups—those with intense $m/z$ 182 ions and those with intense $m/z$ 210 ions. By comparison with the spectra of other cocaine derivatives, it is easy to see that the first group are aroyl esters of methylecgonine ($R^2 = CH_3 = Me$, using the notation in Section 8.3), while the latter group comprises aroyl esters of propylecgonine ($R^2 = CH_3CH_2CH_2 =$ Pr). This assignment is corroborated by first losses from the molecular ions of 31 daltons (MeO·) for methylecgonyl esters and 59 daltons (PrO·) for propylecgonyl esters. (Some of these spectra are so weak and filled with background peaks that identifying the first loss from the molecular ion is difficult.)

Since the intensities of the $m/z$ 82, 182, and 210 ions mitigate against additional substitution anywhere else but the aromatic ring, it remains to determine the nature of this substitution. The mass spectra of A and B have fairly intense $m/z$ 121 ions, while $m/z$ 151 ions are observed in the remaining spectra. Ions at $m/z$ 121 are important in the spectra of derivatives of hydroxycocaine, whereas $m/z$ 151 ions are seen in the mass spectra of the hydroxymethoxycocaines (Table 8.1). Indeed, the molecular ion of compound A occurs at a mass 42 daltons higher than that of hydroxycocaine, corresponding to the derivatization of the aryl $-OH$ group with a propyl group. The molecular ion of compound B is 28 daltons higher still, since this is an ester of propyl-, not methylecgonine. Compound A thus seems to have arisen from propylation of hydroxycocaine,

A - arylpropyloxycocaine

whereas compound B came from propylation of hydroxybenzoylecgonine:

(9.40)

Hydroxybenzoylecgonine          B - arylpropyloxypropylecgonine

The presence of the propyloxy group on the aromatic ring is corroborated in each of these compounds by the presence of a relatively intense $m/z$ 163 peak in the mass spectra, corresponding to the propyloxybenzoyl ion. This ion loses propylene (42 daltons) via a four-center hydrogen rearrangement to give the hydroxybenzoyl ion at $m/z$ 121 [compare Eq. (5.21)]:

$$-CH_3CH=CH_2 \qquad (9.41)$$

m/z 163                    m/z 121

Of the remaining three compounds, C is a derivative of methylecgonine, while the other two are propylated esters of benzoylecgonine derivatives. The $m/z$ 151 and 168 combination from D is highly suggestive of the arylhydroxymethoxycocaines (Table 8.1); in fact, the molecular weight of 377 is 28 daltons higher than that of the hydroxymethoxycocaines, consistent with the alkyl ester group being propyl, rather than methyl. Compound E has a molecular weight 42 daltons higher still, indicating not only propylation of the carboxylic acid group, but also of the aryl hydroxy group. An isomeric methylecgonyl structure having two arylpropyloxy groups (which, from a metabolic standpoint, would have been far more interesting) is inconsistent with the ion at $m/z$ 210. Corroborating the presence of the arylpropyloxymethoxy group in compound E is the appearance of the corresponding benzoyl ion at $m/z$ 193 (42 daltons higher than that shown by compound D). Like the propyloxybenzoyl ion above, it also loses propylene via rearrangement to give the hydroxymethoxybenzoyl ion at $m/z$ 151.

Compound C has been left until last because of the poor quality of its spectrum. The fact that it is a derivative of methylecgonine and also has a molecular weight 42 daltons higher than that of the hydroxymethoxycocaines strongly suggests that it is an arylpropyloxymethoxycocaine. Consistent with this is the peak at $m/z$ 151 (the hydroxymethoxybenzoyl ion; the peak at $m/z$ 193 corresponding to the propyloxymethoxybenzoyl ion is lost in the background clutter in that area of the spectrum). Beyond this, it is difficult to make further comments about this spectrum.

C - arylpropyloxymethoxy
methylecgonine

D - arylhydroxymethoxy
propylecgonine

E - arylpropyloxymethoxy
propylecgonine

## REFERENCES

R.M. Smith, *J. Forensic Sci.,* **42,** 608–616 (1997).
R.M. Smith and L.A. Nelsen, *J. Forensic Sci.,* **26,** 280 (1991).

# INDEX

Pages containing **representations of mass spectra** for individual compounds are shown in **boldface**; those having *interpretations of mass spectra* for individual compounds are shown in *italics*.